AF767175

Redox

Edited by Rozina Khattak

Published in London, United Kingdom

IntechOpen

Supporting open minds since 2005

Redox
http://dx.doi.org/10.5772/intechopen.77556
Edited by Rozina Khattak

Contributors
Melahat Goktas, Antonio Ibarz, Sergio Sánchez-Nuño, Teresa Carbonell, Viorica Parvulescu, Qian Xu, Chunzhen Yang, Huaneng Su, Jose Zagal, Ingrid Ponce, Ruben Oñate, Rozina Khattak, Murtaza Sayed, Muhammad Sufaid Khan, Hamsa Noreen

© The Editor(s) and the Author(s) 2020
The rights of the editor(s) and the author(s) have been asserted in accordance with the Copyright, Designs and Patents Act 1988. All rights to the book as a whole are reserved by INTECHOPEN LIMITED. The book as a whole (compilation) cannot be reproduced, distributed or used for commercial or non-commercial purposes without INTECHOPEN LIMITED's written permission. Enquiries concerning the use of the book should be directed to INTECHOPEN LIMITED rights and permissions department (permissions@intechopen.com).
Violations are liable to prosecution under the governing Copyright Law.

CC BY

Individual chapters of this publication are distributed under the terms of the Creative Commons Attribution 3.0 Unported License which permits commercial use, distribution and reproduction of the individual chapters, provided the original author(s) and source publication are appropriately acknowledged. If so indicated, certain images may not be included under the Creative Commons license. In such cases users will need to obtain permission from the license holder to reproduce the material. More details and guidelines concerning content reuse and adaptation can be found at http://www.intechopen.com/copyright-policy.html.

Notice
Statements and opinions expressed in the chapters are these of the individual contributors and not necessarily those of the editors or publisher. No responsibility is accepted for the accuracy of information contained in the published chapters. The publisher assumes no responsibility for any damage or injury to persons or property arising out of the use of any materials, instructions, methods or ideas contained in the book.

First published in London, United Kingdom, 2020 by IntechOpen
IntechOpen is the global imprint of INTECHOPEN LIMITED, registered in England and Wales, registration number: 11086078, 7th floor, 10 Lower Thames Street, London, EC3R 6AF, United Kingdom
Printed in Croatia

British Library Cataloguing-in-Publication Data
A catalogue record for this book is available from the British Library

Additional hard and PDF copies can be obtained from orders@intechopen.com

Redox
Edited by Rozina Khattak
p. cm.
Print ISBN 978-1-78984-886-1
Online ISBN 978-1-78984-887-8
eBook (PDF) ISBN 978-1-78985-209-7

For EU product safety concerns:
IN TECH d.o.o., Prolaz Marije Krucifikse Kozulić 3, 51000 Rijeka, Croatia, info@intechopen.com or visit our website at intechopen.com.

We are IntechOpen,
the world's leading publisher of
Open Access books
Built by scientists, for scientists

4,900+
Open access books available

123,000+
International authors and editors

140M+
Downloads

151
Countries delivered to

Our authors are among the
Top 1%
most cited scientists

12.2%
Contributors from top 500 universities

WEB OF SCIENCE™

Selection of our books indexed in the Book Citation Index
in Web of Science™ Core Collection (BKCI)

Interested in publishing with us?
Contact book.department@intechopen.com

Numbers displayed above are based on latest data collected.
For more information visit www.intechopen.com

Meet the editor

Asst. Prof. Dr Rozina Khattak serves as the Head of the Department of Chemistry at the Shaheed Benazir Bhutto Women University, Peshawar, Pakistan. She is actively engaged in teaching, administration, research, and publication with several publications in reputed international journals, books, and conference proceedings of national and international level. She completed her PhD degree from the University of Karachi, Department of Chemistry, Pakistan during 2012. She has been a visiting research scholar at the Durham University, Department of Chemistry, University Science Laboratories, Durham-UK during 2009. She joined the National Center of Excellence in Physical Chemistry, University of Peshawar, Pakistan and served as an Assistant Professor for a year, soon after completing her PhD from April 2012 onwards. Later on, she moved to the Shaheed Benazir Bhutto Women University Peshawar, Pakistan and started serving as an Assistant Professor with an additional charge of the Head of Department since October 2014.

Contents

Preface

The book "Redox" provides vast insight into the oxidation-reduction reactions to its readers. The book will help the reader to understand the role of redox potential in reactions; applications of redox reactions in energy production and coordination chemistry, synthesis of polymers, degradation processes of toxic compounds in water treatment, and the impact of its balance in fish welfare. The book consists of three sections that include redox in the coordination compounds, organic compounds and polymerization; redox in electrochemistry; and redox and fish welfare. The first section consists of three chapters that describe the role of redox reactions in several fields such as transition metal chemistry, degradation processes of toxic compounds and dyes in treatment of water and wastewater, the catalysis of oxidation of organic compounds by metal active sites and synthesis of copolymers. The role of redox reactions and reactivity description of compounds are discussed in the second section of the book. This section consists of two chapters. The non-aqueous redox flow batteries have been described in this section. The third section extensively discusses the redox balance and fish welfare and consists of one chapter. The book provides a good contribution to understand redox chemistry and its applications in a variety of processes of chemical, industrial, and environmental concerns. The editor acknowledges the efforts and hard work of all the authors and their co-authors for their worthy contributions. The editor also acknowledges the help and patience of the author service manager, Ms. Romina Skomersic for her guidance and assistance. This book will hopefully be a significant contribution on the topic of "Redox".

Dr. Rozina Khattak
Assistant Professor,
Head of Department,
Department of Chemistry,
Shaheed Benazir Bhutto Women University,
Peshawar, Pakistan

Section 1

Redox in Coordination Compounds/Organic Compounds and Polymerization

Chapter 1

Introductory Chapter: Redox - An Overview

Rozina Khattak, Murtaza Sayed, Muhammad Sufaid Khan and Hamsa Noreen

1. Redox

The term "redox" is the combination of two different terms that describe two totally different chemical processes, i.e., "reduction" and "oxidation." The abbreviation "red" has been taken to distinguish reduction from oxidation that is "ox." The reduction is a process wherein any chemical entity gets reduced. It is different from oxidation, which is opposite of the reduction because the chemical entities are oxidized. These two processes or reactions which simultaneously take place in a system are abbreviated as "redox."

1.1 General introduction

Oxidation-reduction was primarily used to describe the reaction(s) of combination and/or removal of oxygen with or from chemical substances, respectively. Simultaneously, the removal and/or the addition of hydrogen were also used to differentiate among oxidation and reduction, respectively. The definitions were extended to a broader level, and the changes in the oxidation number or oxidation state of elements were considered to define oxidation and reduction. The increase in the oxidation number leads to oxidation and its alternative process yields reduction. This vast definition encompasses the recent and exact interpretation of "redox" reactions that is acceptance and donation of the electron(s) between the reacting entities. Consequently, the redox phenomenon indicates a simple reaction, formation of carbon dioxide as a consequence of the oxidation of carbon and/or formation of methane by the reduction of carbon, for example, and the complex reaction consisting of a number of electron transfer reactions during the oxidation of sugar in the human body to produce energy.

The redox reaction(s) involves an oxidant or oxidizing agent and a reductant or reducing agent. The oxidant takes the electron(s) and oxidizes the reductant. The reductant, however, donates the electron(s) and reduces the oxidant.

Redox reactions are the key to make many desired chemical changes and/or processes reality to get maximum benefit out of it. A simple overview to surface the vital need of these reactions revolves around combustion, metabolic reactions, extraction of metals from their ores, manufacture of countless chemicals, and reactions occurring in our natural environment. For example, a cell either battery and/or biological cell involves redox processes [1–3]. Research that involves the biological systems interprets that the electronic and the structural environment of the substance(s) are the key factors that control chemical transformations such as electron transfer mechanisms in DNA molecules, which may appear through the exposure of cells to radiations that may have the power of ionization to cause biologically deleterious effects such as inactivation, transformation, and mutagenesis [4–6]. The water molecules are ionized by some specific radiations and form radicals in the vicinity of DNA that contribute to the significant damage in DNA and chemical modifications to DNA, consequently. Of these radicals, the hydroxyl radical is thought to be the most damaging and produces the consequences for DNA strand breakage by the redox dependence of the rate of interaction of hydroxyl radical adducts of DNA nucleobases with oxidants [7].

Redox reactions mainly follow second-order kinetics with a series of intermediate reactions in a range of mechanisms such as entity transfer mechanisms that involve electron, atom, or group transfer and ligand addition, substitution, or dissociation.

Redox phenomenon in terms of electron transfer (ET) reactions and their mechanisms is catered for the interest of readers of this book. Essentials of only the electron transfer reactions of coordination or transition metal complexes and advanced oxidation processes for water treatment are being focused in a brief and narrative way.

1.2 Redox reactions of transition metal complexes

An enormous number of studies unfolded the characteristics and effect of the structural geometries on the kinetics and mechanisms of the redox reactions of various transition metal complexes [8–25]. The literature review helped to summarize that the redox reactions of the transition metal complexes undergo two types of mechanisms. These types are classified as the *outer-sphere and inner-sphere mechanisms*, which lead the electron transfer processes of the transition metal complexes.

1.3 Schematic representation of the mechanistic pathways

The fundamental distinction between the two mechanistic routes of electron transfer is the simplicity of the outer-sphere mechanism over the inner-sphere mechanism. The outer-sphere redox reactions are simple in nature and undergo electron transfer in a very simple way. The outer-sphere mechanism is further classified into the self-exchange and cross-exchange reactions. This classification is based upon the oxidation state of coordination compounds. In the self-exchange reactions, the same coordination compounds with different oxidation states reduce and oxidize each other. However, in the cross-exchange reactions, different coordination compounds with either of the same and/or different oxidation states or numbers reduce and oxidize each other in the vicinity. However, in the inner-sphere mechanism of electron transfer, the substitution of ligand or atom prior to electron transfer plays a key role. The difference between the two mechanisms is represented in **Figure 1** [13].

One cannot easily propose the operated mechanism of electron transfer under specific cases beside the simple and apparent difference between the two reaction pathways. There are two reasons for this. It may usually be possible to suppose without any doubt that the inner-sphere mechanism is operating the electron transfer process in favorable cases, but in many reactions where the reactants or the products and/or both of them are substitution labile, the mechanism through the inner-sphere process becomes suspected. In such reactions, the exact nature of the real reacting entities that are taking part in the reactions and the products which form initially becomes impossible to recognize without proper experimental and technical treatments. The other reason for ambiguity in recognizing the reliable electron transfer mechanism appears when the nature of the outer-sphere mechanism is considered, which does not need any re-arrangement of the structure of the reacting entities rather it only needs the transfer of an electron between the oxidizing

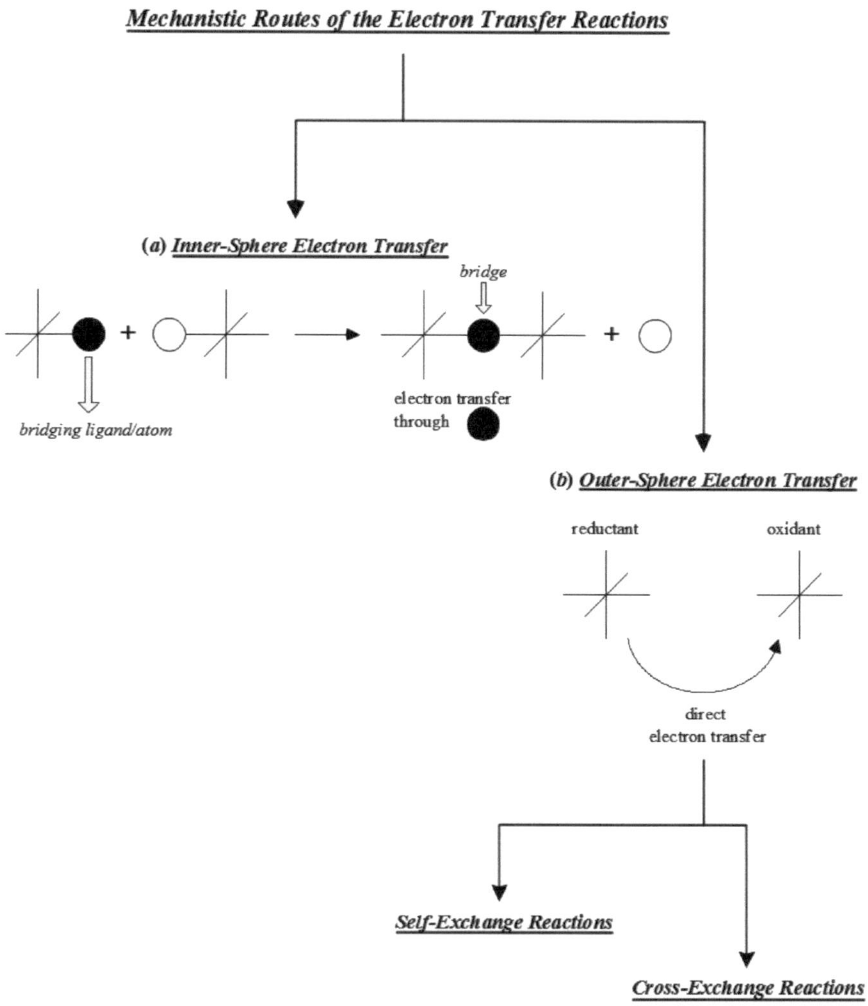

Figure 1.
Schematic representation of the redox mechanism. (a) The inner-sphere mechanism. (b) The outer-sphere mechanism.

and the reducing agents. To suggest or propose an outer-sphere mechanism one needs credible evidence with proof of unavailability of the alternative inner-sphere mechanistic pathway. Consequently, there are a large number of reactions that are clearly defined to be operated through inner-sphere mechanism. However, many reactions follow outer-sphere mechanism and an uncomfortably big number of reactions operates by inner-/outer-sphere mechanism i.e., in between [26].

1.4 Experimental approach: kinetics and mechanisms of some selected transition metal complexes of Fe(II) and Fe(III)

The redox reactions of a few selected coordination compounds of the transition metal, iron, in its two oxidation states, i.e., +2 and +3, are briefly discussed. The mixed ligand complexes such as dicyanobis(phenanthroline)iron(III) and dicyanobis(bipyridine)iron(III) oxidize hexacyanoferrate(II), acetylferrocene, and 1-ferrocenylethanol by outer-sphere mechanism [8–11, 13] in the aqueous-organic media. The effect of optimized parameters on the kinetics of the redox reactions helped to propose the operated mechanism and rate laws (**Figures 2–5**) [8, 10].

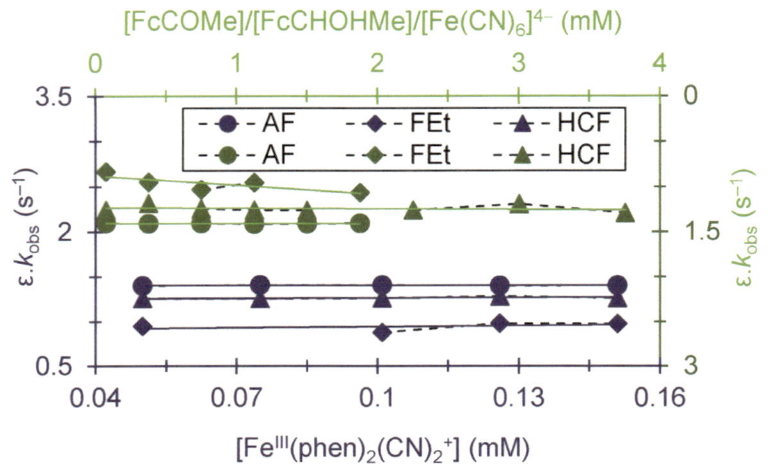

Figure 2.
Kinetics of the redox reaction between Fe(III) and Fe(II) complexes. The abbreviations; $\varepsilon.K_{obs}$/AF/FEt/HCF, correspond to the multiplication product of the molar absorptivity of $[Fe^{II}(phen)_2(CN)_2]$ and observed zero-order rate constant/acetylferrocene/1-ferrocenylethanol/hexacyanoferrate(II).

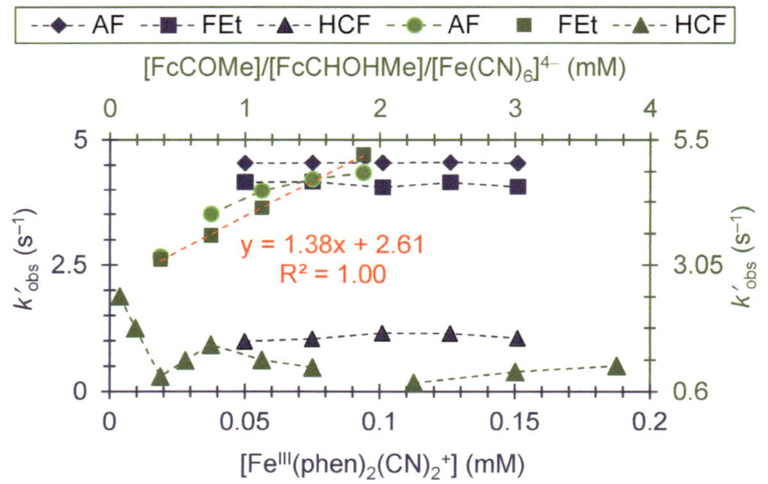

Figure 3.
Kinetics of the redox reaction between Fe(III) and Fe(II) complexes. The abbreviations; k'_{obs}/AF/FEt/HCF, correspond to the observed pseudo-first-order rate constant/acetylferrocene/1-ferrocenylethanol/hexacyanoferrate(II).

An outer-sphere electron transfer mechanism was proposed for each reaction because the oxidants and reductants are substitution inert and outer-sphere reactants.

Meanwhile, the oxidation of tris(bipyridine)iron(II) by ceric and bromate ions in aqueous-acidic media was reported to follow an outer-sphere mechanism with a second-order rate law (**Figures 6** and 7) [12, 14]. The iron complexes of the chelating agents such as 1,10-phenanthroline and 2,2′-bipyridine and the ligand such as cyanide ion are either good outer-sphere oxidants and/or reductants and show high stability towards ligand substitution [27, 28].

Protein as a nutrient is an important structure and is critical to aerobic life because of its control of oxygen reduction reactions. This management is crucial either to avoid or to minimize the production of destructive products such as hydroxyl radicals, peroxide, and superoxide as a result of the destructive reduction

$$Rate = k_1 + \frac{k_2 K_{eq}\left[Fe^{III}(phen)_2(CN)_2^+\right][FcCOMe]_T\left[H^+\right]}{1 + K_{eq}\left[H^+\right]}$$

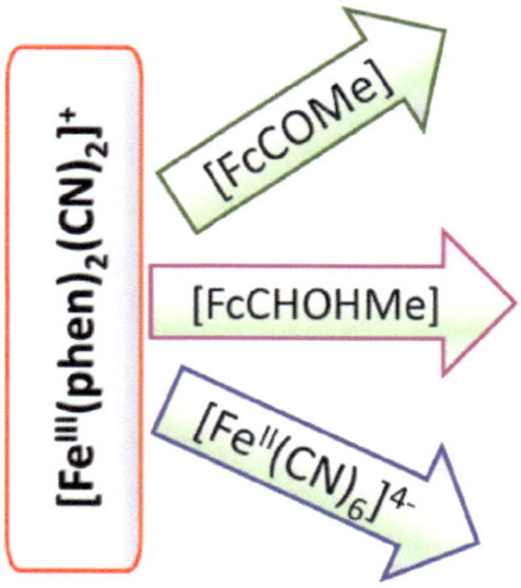

$$Rate = k_1 + \frac{k_2\left[Fe^{III}(phen)_2(CN)_2^+\right][FcCHOHMe]_T}{1 + K_{eq}\left[H^+\right]}$$

$$Rate = k_1 + \frac{k_2\left[Fe^{III}(phen)_2(CN)_2^+\right]\left[Fe^{II}(CN)_6^{4-}\right]_T}{\left(1 + K_1\left[H^+\right]\right)} + \frac{k_3\left[Fe^{III}(phen)_2(CN)_2^+\right]\left(1 + K_2\left[H^+\right]\right)}{K_2\left[H^+\right]\left[Fe^{II}(CN)_6^{4-}\right]_T}$$

Figure 4.
Proposed rate laws: reduction of $[Fe^{III}(phen)_2(CN)_2]^+$ by acetylferrocene, 1-ferrocenylethanol, and hexacyanoferrate(II) [8].

$$-\frac{d\left[Fe(II)\right]}{dt} = k_1 + k_2\left[Fe(III)\right]\left[Fe(II)\right]_T + \frac{k_3\left[Fe(III)\right]}{K_2\left[H^+\right]\left[Fe(II)\right]_T}$$

Figure 5.
Proposed rate law: oxidation of $[Fe^{II}(CN)_6]^{4-}$ by $[Fe^{III}(phen/bpy)_2(CN)_2]^+$ [10].

$$-\frac{d\left[Fe^{II}(bpy)_3^{2+}\right]}{dt} = \frac{k K_1 K_d \left[Fe^{II}(bpy)_3^{2+}\right][Ce^{IV}]_T\left[SO_4^{2-}\right]_a}{K_d^2 + \left[SO_4^{2-}\right]_a\left[K_d\left(K_1 + [H^+]_a\{K_2 + K_1K_3\}\right) + \left(K_4[H^+]_a\left[SO_4^{2-}\right]_a\right)(K_2 + K_1K_3)\}\right]}$$

Figure 6.
Proposed rate law: oxidation of $[Fe^{II}(bpy)_3]^{2+}$ by ceric sulfate in the aqueous-acidic media [13–14].

$$Rate = k_1\left[Fe(2,2'-bipy)_3^{2+}\right][HBrO_3] + \frac{k_2 K_{ip}\left[Fe(2,2'-bipy)_3^{2+}\right][H_2BrO_3^+]}{(1 + K_{ip}[H_2BrO_3^+])}$$

Figure 7.
Proposed rate law: oxidation of $[Fe^{II}(2,2'\text{-}bipy)_3]^{2+}$ by bromate ion in the aqueous-acidic media [12].

and/or to optimize the utilization of oxygen in an effective way to transport and storage. Hemeproteins are those structures which maintain and control these oxygen reduction reactions. Such management and control of protein was surfaced by studying the mechanism of the redox reaction of aquopentacyanoferrate(II) ($[Fe^{II}(CN)_5H_2O]^{3-}$) with coordinated dioxygen of human oxyhemoglobins (HbO_2) [29]. This reaction yielded hydrogen peroxide (H_2O_2) and aquomethemoglobin ($metHb{\bullet}H_2O$) and the oxidation product of aquopentacyanoferrate(II), i.e., $[Fe^{III}(CN)_5H_2O]^{2-}$. The reaction was found to undergo overall second-order kinetics with a first order in each oxidant (HbO_2) and reductant ($[Fe^{II}(CN)_5H_2O]^{3-}$), respectively. The results declared that the structures of the reactants such as protein and external donor control the kinetics of the electron transfer with an inner-sphere mechanism that involves direct electron transfer from the aquopentacyanoferrate(II) to bound dioxygen that yields peroxide, subsequently. Another study surfaced the effect of binding sites and protonation on the kinetics of the electron transfer reaction (s) of blue copper proteins [30]. The oxidants with different binding sites such as $[Co^{III}(4,7\text{-DPSphen})_3]^{3-}$, $[Fe^{III}(CN)_6]^{3-}$, and $[Co^{III}(phen)_3]^{3+}$ were used to oxidize parsley plastocyanin. In each reaction, regardless of the binding sites, and prior to electron transfer, a strong association between protein and complex occurs. The variation in the binding sites varied the reduction potential and affected the rate of electron transfer, consequently. The reductant (plastocyanin) is a copper protein that consisted of type 1 copper, which is involved in electron transport from photosystem II to photosystem I at the surface of the thylakoid membrane. A single copper here utilizes oxidation states I and II. The structure of poplar plastocyanin PCu^{II} contains Cu(II) coordinated with two histidines, one cysteine, and one methionine in a distorted tetrahedral arrangement.

It has always been of interest to probe the details of the transfer of electron(s) and proton(s) because of successfully unveiling strategies of energy conversion in both of the fields, biology and chemistry. The energies as well as mechanism are strongly influenced by the coupling of electron and proton transfer. This defines the need to build up multiple redox equivalents to carry out those reactions that involve multielectrons. This also explains those mechanistic pathways through which electron and proton transfer occur simultaneously to avoid intermediates of high energy [31]. The theoretical background of the proton-coupled electron transfer reactions in solutions and proteins and electrochemistry was reviewed and discussed [32]. The theoretical treatment was based on the calculations of multistate continuum theory wherein the solvent provides dielectric continuum, the solute is treated as a multistate valence bond model, and quantum mechanical approach is used for transferred proton or hydrogen nucleus. The rate expression of electronically nonadiabatic electron transfer and proton-coupled electron transfer depends upon the reorganization energies of solute (inner-sphere) and solvent (outer-sphere) and also upon electronic coupling. For proton-coupled electron transfer, this is the average of the proton vibrational wave functions of the reactants and the products. The compensation of the smaller outer-sphere solvent reorganization energy for proton-coupled electron transfer by the larger energy needed to coupling for electron transfer appears with a similar rate for both electron transfer and proton-coupled electron transfer in calculations. A comparative theoretical study supported the reviewed outcomes through the proton-coupled electron transfer, single proton transfer, and single electron transfer reactions in iron bi-imidazoline complexes [33].

1.5 Advanced oxidation processes for water treatment

The oxidation of organic compounds by a number of oxidants either of inorganic nature or organic nature has been of interest. These redox reactions are usually catalyzed by transition metals. The kinetics of the oxidation of pyridinecarbaldehyde isonicotinoyl hydrazone to isonicotinoyl picolinoyl hydrazine was studied, and the mechanism was proposed in the view of results obtained in aqueous solution [34]. The reaction was catalyzed by iron(III). Advanced oxidation processes (AOPs) are used to remove pollutants/contaminants such as organic and inorganic compounds from water and wastewater by oxidation of these unwanted compounds. The process involves a number of chemical reactions consisting of oxidants such as ozone (O_3), hydrogen peroxide (H_2O_2), and UV light with the addition of catalysts that may lead to yield hydroxyl radical ($^{\bullet}OH$) which degrades such pollutants, dyes and organic compounds, for example [35–40].

2. Conclusion

This concise review of the redox reactions and their applications surfaced the crucial role of redox processes. The importance of redox processes is undoubtedly tremendous. The applications encompass energy production, technological development to treat and maintain water resources, and advances in materials chemistry. These advances may lead the life to its standard in an economic and cost-effective way. Redox reactions are also an important facet of biological and biochemical world to carry out life and its routine practices. For example photosynthesis, respiration and digestion are among the common ones. Precisely, we can sum up with one sentence that "redox" is basically the key to sustaining life on this planet.

Author details

Rozina Khattak[1*], Murtaza Sayed[2], Muhammad Sufaid Khan[3] and Hamsa Noreen[4,5]

1 Department of Chemistry, Shaheed Benazir Bhutto Women University, Peshawar, Pakistan

2 National Center of Excellence in Physical Chemistry, University of Peshawar, Peshawar, Pakistan

3 Department of Chemistry, University of Malakand, Chakdara Dir Lower, Pakistan

4 Institute of Chemical Sciences, University of Peshawar, Peshawar, Pakistan

5 Department of Physical Chemistry, Institute of Chemistry, University of Campinas, Campinas, São Paulo, Brazil

*Address all correspondence to: rznkhattak@yahoo.com; rznkhattak@sbbwu.edu.pk

IntechOpen

© 2020 The Author(s). Licensee IntechOpen. This chapter is distributed under the terms of the Creative Commons Attribution License (http://creativecommons.org/licenses/by/3.0), which permits unrestricted use, distribution, and reproduction in any medium, provided the original work is properly cited.

References

[1] Matsuda Y, Tanaka K, Okada M, Takasu Y, Morita M, Matsumura-Inoue T. A rechargeable redox battery utilizing ruthenium complexes with nonaqueous organic electrolyte. Journal of Applied Electrochemistry. 1988;**18**:909-914

[2] Ramirez BE, Malmström BG, Winkler JR, Gray HB. The currents of life: The terminal electron-transfer complex of respiration. Proceedings of the National Academy of Sciences of the United States of America. 1995;**92**: 11949-11951

[3] Brzezinski P, Larsson G. Redox-driven proton pumping by heme-copper oxidases. Biochimica et Biophysica Acta - Bioenergetics. 1605;**2003**:1-13

[4] Giese B. Electron transfer in DNA. Current Opinion in Chemical Biology. 2002;**6**:612-618

[5] Núñez ME, Barton JK. Probing DNA charge transport with metallointercalators. Current Opinion in Chemical Biology. 2000;**4**:199-206

[6] Jackson NM, Hill MG. Electrochemistry at DNA-modified surfaces: New probes for charge transport through the double helix. Current Opinion in Chemical Biology. 2001;**5**:209-215

[7] Bamatraf MMM, O'Neill P, Rao BSM. Redox dependence of the rate of interaction of hydroxyl radical adducts of DNA nucleobases with oxidants: Consequences for DNA strand breakage. Journal of the American Chemical Society. 1998;**120**:11852-11857

[8] Khattak R, Naqvi II. Study of the effect of structure on the kinetics and mechanism of the redox reactions of Fe (III)/Fe(II) complexes. Bulgarian Chemical Communications. 2019;**51**: 129-134

[9] Khattak R, Naqvi II, Summer S, Sayed M. Mechanism of the oxidation of 1-(ferrocenyl)-ethanone/ethanol by dicyanobis(phenanthroline)iron(III). Arabian Journal of Chemistry. 2019;**12**: 4240-4250

[10] Khattak R, Khan MS, Naqvi II. Mechanism of the electron-exchange reactions between mixed ligand Fe(III) complexes and cyano complex of Fe(II). Bulgarian Chemical Communications. 2018;**50**:38-44

[11] Khattak R, Nazir M, Summer S, Sayed M, Minhaz A, Naqvi II. Thermodynamic aspect: kinetics of the reduction of dicyanobis(phen)iron(III) by acetylferrocene and methylferrocenemethanol. Chemical Papers. 2018;**72**:883-893

[12] Summer S, Naqvi II, Khattak R, Gulzar S, Reyaz F. Kinetics and mechanism of $[Fe\ (bipy)_3]^{2+}$ and $[BrO_3^{-}]$ system in aqueous acidic medium. Journal of the Chemical Society of Pakistan. 2016;**38**: 384-389

[13] Khattak R. Comparative kinetic study for the electron transfer reactions of some iron complexes [PhD]. Karachi: University of Karachi; 2011

[14] Khattak R, Naqvi II, Farrukh MA. Kinetics and mechanism of the oxidation of a ferrous complex with an α,α′-diimine chelate ligand by ceric sulfate in aqueous acidic medium by UV-Vis absorption spectroscopy. Journal of the Iranian Chemical Society. 2008;**5**:631-640

[15] Doctorovich F, Granara M, Salvo FD. The reaction of $[Ru\ (bpy)_2(NO)Cl]^{2+}$ and $[Fe(CN)_5NO]^{2-}$ with benzylamine coordinated nitrosyl as an oxidizing agent. Transition Metal Chemistry. 2001;**26**:505-509

[16] Lee D-J, Kim B-G, Kim S-G, Kim D-Y, Lee B-W, Doh M-K. Stereoselectivity of electron transfer reactions (II); reaction of lamda-[Co(EDDS)]$^-$ and Co(II) complexes with stereospecific diamine ligands. Journal of Coordination Chemistry. 1996;**37**: 227-236

[17] Tomiyasu H, Gordon G. Kinetics and mechanism of the oxidation of bis(terpyridine)-iron(II) by ozone in acidic aqueous solutions. Ozone: Science & Engineering. 1989;**11**:59-67

[18] Ibrahim MS, Gemeay AH, Etaiw SE-dH. Oxidation of a three-dimensional polymeric iron(II) complex with sodium nitrite in acidic medium. Transition Metal Chemistry. 2001;**26**: 44-49

[19] Abdel-Khalek AA, Hassan ES, Mohamed RA. Mechanism of electron transfer reactions of ternary nitrilotriacetato cobaltate(II) complexes involving maleate and tartarate by periodate. Journal of Coordination Chemistry. 2008;**61**:152-166

[20] Davies KM, Espenson JH. Electron-transfer reactions of vanadium(II) and some monosubstituted pentacyanocobaltate(III) ions. Journal of the American Chemical Society. 1969; **91**:3093-3094

[21] Cavasino FP, Sbriziolo C, Turco Liveri ML. Kinetic study of the iron(III) oxidation of some alkyl-substituted ferrocenes in water-in-oil microemulsions of AOT in heptane, isooctane and decane. Journal of the Chemical Society, Faraday Transactions. 1998;**94**:395-399

[22] Sachinidis JI, Shalders RD, Tregloan PA. Measurement of redox reaction volumes for iron(III/II) complexes using high-pressure cyclic staircase voltammetry. Half-cell contributions to redox reaction volumes. Inorganic Chemistry. 1994;**33**: 6180-6186

[23] Ige J, Ojo JF, Olubuyide O. Kinetics and mechanism of oxidation of tris-(1,10-phenanthroline)iron(II) by chlorine and bromine and of the reduction of tris-(1,10-phenanthroline) iron(III) by iodide ions. Canadian Journal of Chemistry. 1979;**57**: 2065-2070

[24] Mukherjee K, Mukherjee DC, Moulik SP. Reaction kinetics in a microemulsion medium IV. Hexacyanoferrate(III)-iodide reaction in water/aerosol-OT/heptane microemulsion and mixed solvents. Bulletin of the Chemical Society of Japan. 1997;**70**:1245-1253

[25] Partha Sarathi TVN, Kalyan Kumar A, Krishna Kishore K, Vani P. Kinetics and mechanism of oxidation of glycine by iron(III)–1,10-phenanthroline complex in perchloric acid medium. Journal of Chemical Sciences. 2005;**117**:329-332

[26] Tobe ML, Burgess J. Inorganic Reaction Mechanisms. Longman, New York: Addison Wesley Longman Limited; 1999. pp. 376-524

[27] Khattak R, Naqvi II. Addition product of iron(II) complex of aromatic diimine with sulphuric acid. Journal of Research (Science). 2007;**18**: 219-235

[28] Khattak R, Naqvi II. Identification, preparation and characterization of some iron complexes. Journal of Research (Science). 2008;**19**:17-35

[29] Kawanishi S, Caughey WS. Mechanism of electron transfer to coordinated dioxygen of oxyhemoglobins to yield peroxide and methemoglobin. Protein control of electron donation by aquopentacyanoferrate(II). The Journal

of Biological Chemistry. 1985;**260**: 4622-4631

[30] Lappin AG, Segal MG, Weatherburn DC, Sykes AG. Kinetic studies on 1:1 electron-transfer reactions involving blue copper proteins. 2. Protonation effects and different binding sites in the oxidation of parsley plastocyanin with $Co(4,7\text{-}DPSphen)_3^{3-}$, $Fe(CN)_6^{3-}$, and $Co(phen)_3^{3+}$. Journal of the American Chemical Society. 1979; **101**:2297-2301

[31] Huynh MHV, Meyer TJ. Proton-coupled electron transfer. Chemical Reviews. 2007;**107**:5004-5064

[32] Hammes-Schiffer S, Soudackov AV. Proton-coupled electron transfer in solution, proteins, and electrochemistry. The Journal of Physical Chemistry. B. 2008;**112**:14108-14123

[33] Iordanova N, Decornez H, Hammes-Schiffer S. Theoretical study of electron, proton, and proton-coupled electron transfer in iron bi-imidazoline complexes. Journal of the American Chemical Society. 2001;**123**:3723-3733

[34] Bernhardt PV, Martínez M. The Fe-catalyzed oxidation of aroyl hydrazones to aroyl hydrazines: Mechanistic insight to a remarkable reaction. Journal of Coordination Chemistry. 2010;**63**: 2619-2628

[35] Rehman F, Sayed M, Khan Javed A, Shah Luqman A, Shah Noor S, Khan Hasan M, et al. Degradation of crystal violet dye by Fenton and photo-Fenton oxidation processes. Zeitschrift für Physikalische Chemie. 2018;**232**(12): 1771-1786

[36] Sayed M, Shah LA, Khan JA, Shah NS, Khattak R, Khan HM. Competition kinetics: An experimental approach. In: Farrukh MA, editor. Advanced Chemical Kinetics. Rijeka: InTech; 2018. p. Ch. 05

[37] Ismail M, Khan HM, Sayed M, Cooper WJ. Advanced oxidation for the treatment of chlorpyrifos in aqueous solution. Chemosphere. 2013;**93**:645-651

[38] Sayed M, Ismail M, Khan S, Tabassum S, Khan HM. Degradation of ciprofloxacin in water by advanced oxidation process: Kinetics study, influencing parameters and degradation pathways. Environmental Technology. 2016;**37**:590-602

[39] Khan J, Sayed M, Ali F, Khan Hasan M. Removal of acid yellow 17 dye by Fenton oxidation process. Zeitschrift für Physikalische Chemie. 2018;**232** (04):507-525

[40] Khan S, Sayed M, Sohail M, Shah LA, Raja MA. Chapter 6. Advanced oxidation and reduction processes. In: Ahuja S, editor. Advances in Water Purification Techniques. Elsevier; 2019. pp. 135-164. Available from: DOI: 10.1016/B978-0-12-814790-0.00006-5

Chapter 2

Catalytic Behavior of Metal Active Sites From Modified Mesoporous Silicas in Oxidation of Organic Compounds

Viorica Parvulescu

Abstract

The ordered mesoporous silicas containing transition metals are versatile catalytic materials for oxidation of a wide range of organic compounds. In order to obtain active catalysts, different active redox metal sites have been introduced into specific locations (mesoporous channels and framework) of ordered mesoporous silicas (OMSs). All the reported results evidenced that localization of metal ions, their interaction with another metal (bimetallic catalysts) and the support, with typical properties of the ordered mesoporous silica, influenced the oxidation state, respectively their redox properties. To support this, the results regarding the specific properties of transitional metals in the ordered structure of silica, obtained using various characterization methods, were presented. The activity of metal sites in the oxidation reactions was evidenced in various applications carried out in the liquid or gaseous phase. The oxidation of various organic compounds in liquid phase with H_2O_2 (especially aromatic compounds and of alcohols) and in gas phase with oxygen from air was presented for a variety of metal-modified OMSs. The active sites and possible reaction mechanisms were presented for some catalytic systems. The immobilization of multiple metal active species on new ordered porous frameworks and their synergistic interactions remain topics of interest in the future.

Keywords: redox metal sites, transition metals, mesoporous silicas, oxidation reactions

1. Introduction

The ordered mesoporous silicas (OMSs) containing transition metals are versatile catalytic materials for oxidation of a wide range of organic compounds [1–3]. Mesoporous materials with various structures such as MCM-41 (2d hexagonal p6mm) [4–6], MCM-48 (3d cubic Ia3d) [7, 8], SBA-15 (large-pore 2d hexagonal p6mm) [2, 9–12], SBA-16 (large-pore 3d cubic Im3m) [13], KIT-5 (well-ordered cage-type mesoporous with cubic Fm3m) [2], and KIT-6 (large-pore cubic with interpenetrated cylindrical mesopores Ia3d) [2, 14–18] have been used as supports for transition metals. These supports with controlled morphology and accessible metallic center for the reactant molecules also offered the opportunity to immobilize the transition metal complexes and their heterogenization [19, 20]. These catalysts have attracted much interest due to the desirable characteristics of the silica supports such as narrow pore size, high surface area and large pore volume, tunable mesoporous channels with well-defined pore-size distribution, controllable wall composition, and modifiable surface properties. Pore diameter of mesoporous silicas (2–50 nm) and porous structure are usually tailored by the choice of the template surfactant or the incorporation of swelling agents to expand the surfactant micelles during synthesis [21, 22]. In condition of typical synthesis environment of the mesoporous molecular supports, the incorporation of metal methods varies with properties of their precursors.

In order to obtain active catalysts, different active redox metal sites have been introduced into specific locations (mesoporous channels and framework) of the OMSs supports by direct synthesis methods (framework substitution) or post-synthetic methods. In any case, Men^+ can be simultaneously present in different coordination geometries and positions (surface, lattice) [16, 23]. In a direct-synthesis preparation, the condensations of silicon and metal species around the organic micelles occur simultaneously, and it is likely that some of the metal species are trapped in the silica walls during the formation of OMSs supports. This may influence the unit cell parameters, the wall thickness, and the long-range ordering of the material. By contrast, metal species introduced by a postsynthesis treatment (template ion exchange, impregnation, grafting, chemical vapor deposition methods) are mostly located at the surface of the mesopores and do not modify the internal composition of the silica walls, mainly when the samples are prepared in alcohol. The synthesis method offers the advantage that the dispersion and location of metal species are easily controlled. This may be a great advantage with respect to conventional synthesis methods to prepare materials with specific applications in catalysis. The activity of the obtained materials was demonstrated in various reactions, mostly oxidation reactions of the organic compounds in the liquid or gaseous phase. All the reported results show that the localization of the metal ion, morphology, particle and pore channel sizes and their interaction with the support, and other metal (bimetallic catalysts) influence the oxidation state of the catalytic sites, respectively their redox properties. Therefore, the redox properties of these materials are the result of the support and metal cation synergistic effect.

2. Location, chemical state, and environment of the incorporated metals

Catalytic properties of the incorporated transition metals in OMSs supports were mainly attributed to their location, chemical state, and environment. The location of metals in mesoporous silica network is the result of the synthesis method, the intrinsic properties of the incorporated metal, and silica support. The location, the loading, and the properties of the incorporated metals can also influence the support mesoporous structure. Thus, the variation of (100) peak in X-ray diffractograms obtained at a lower angle has been detected in many Me-MCM-41 patterns, indicating the effect of metal species on the ordered mesoporous structure of the support [3, 24, 25]. The higher concentrations of metal species ($Me[(OH)n(H_2O)m]$) at the interface affect the electrostatic interaction between surfactant and silica precursor and the polymerization processes of the silica system in alkaline media (pH 11) during synthesis. In such conditions, the free energy of the mesostructure formation decreases and a mixture of nonstructurated oxides (SiO_2–Co_3O_4) was exhibited [25]. Changing the Si-O-Si bond angle due to incorporation of metal in the mesoporous silica support increases the number of local defects within the mesoporous structure. IR spectroscopy is one of the first techniques that has been used for the characterization of materials with metal incorporated in mesoporous silica to obtain information on changes of the Si-OH from surface or regarding appearance of the new vibrations as (Si-O-Me). The change in the Si-O-Si bond angle due to incorporation of Me increases the number of local defects within the mesoporous structure. A much disputed is the change in intensity of the band at 960 cm^{-1} indicating the structural changes in the Si-OH surface due to the presence of metal oxide or to the evolution of new vibrations (Si-O-Me) that appear in the same region [26].

For TiMCM-41 synthesized by surfactant-assisted direct hydrothermal (DHT) method [7, 16, 21], the absence of the characteristic bands for crystalline TiO_2 phase in the wide angle XRD patterns reveal that the metal ions were either atomically dispersed in MCM-41 framework or may exist in an amorphous dispersed form on the outside surface of mesoporous support. Diffuse reflectance UV–Vis spectra also revealed distorted tetrahedral environments for titanium inside the MCM-41 matrix or octahedrally coordinated titanium sites, due to the possible hydration effects. The presence of titanium in various mesoporous silica supports in +4 oxidation state was confirmed by ESR and XPS analysis [7, 9, 23]. XPS analysis was used as an additional tool of UV–VIS data since the dispersion of Ti species depends strongly on the synthesis method, properties of the support, and other metals [7, 9, 16]. XRD and spectroscopic results reveal that the titanium was dispersed as titanium ions on SBA-15 silica wall surfaces at low titanium loading, whereas a titanium dioxide anatase film was formed at high titanium loading [16, 23]. Thus, the nature of Ti species on TiMCM-41 s surface, prepared by three different methods, i.e., isomorphous substitution, wet impregnation, and mechanical mixing, was analyzed by means of Raman spectra [27]. The obtained results indicated the presence of anatase for TiO_2MCM-41 sample obtained by mechanical mixing. In contrast, no Raman-active bands were observed on TiMCM-41, obtained by direct synthesis, indicating the absence of surface anatase phase. For TiMCM-41, obtained by impregnation method, no surface anatase phase was detected at a low Ti/Si ratio of 0.3, while surface anatase was detected at high Ti/Si ratios such as 0.6 and 0.9. UV–Vis spectra showed for TiMCM-41 a strong absorbance band at 210 nm and a shoulder band at 260 nm. The first was attributed to isolated Ti atoms in tetrahedral coordination, while the band at 260 nm was attributed to isolated Ti atoms in pentahedral or octahedral coordination. Therefore, it is believed most Ti atoms should substitute Si in the framework or surface in TiMCM-41 with the formation of Ti-O-Si-O-Ti bands. For TiO_2MCM-41, adsorption bands at 220, 260, and 320 nm were clearly observed in the UV–vis spectra. The band at 320 nm was typical for bulk titania, indicating the existence of bulk TiO_2. The band at 220 was attributed to isolated Ti atoms with distorted tetrahedral environment. These Ti species, dispersed on OMSs support, were supported along with other cations (Ce, V, Nb) or was used as support for another active metal as Ce, Pt, Fe [7, 9, 16, 21, 28]. The second or third [16] metal was evidenced by SEM backscattering and TEM microscopy, as extra framework nanoparticles (**Figure 1**). A good correlation between TEM results (**Figure 1C**) and H_2 chemisorption on Pt nanoparticles' diameter was observed.

XPS spectroscopy sustained the interaction of the second metal with titanium [29] and together on the third metal [16]. The presence of Pt^0 on surface and the effects of titanium loading and of cerium on its percent were explained by metal-support interaction considering TiKIT-6 and CeTiKIT-6 samples as supports for Pt (**Figure 2**). Due to Ti and Ce redox properties and strong interaction with noble metals, these metal oxides influence Pt/PtO molar ratio on the catalyst surface. The extended X-ray absorption fine structure measurements evidenced for Pt immobilized on SBA-15, in absence of Ti and Ce species, the presence of Pt–oxygen chemical bonds at the surface. The concentration of these Pt species increased for Pt-amino-SBA-15 sample [11].

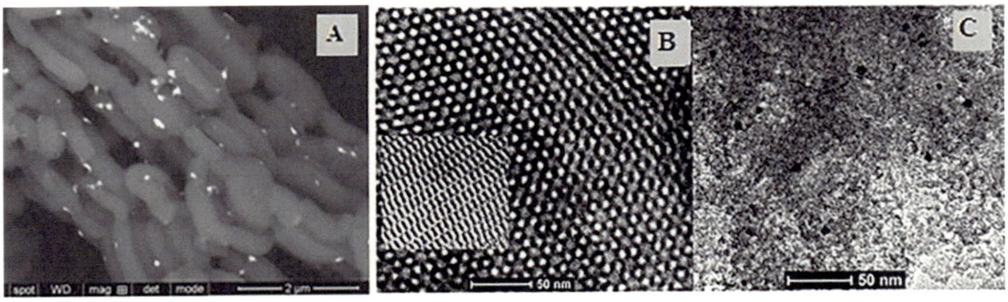

Figure 1.
SEM backscattering (A) of PtTi-SBA-15 (unpublished) and TEM images of TiKIT-6 (B) and PtTiKIT-6 (C) samples (with permission from Ref. [16]).

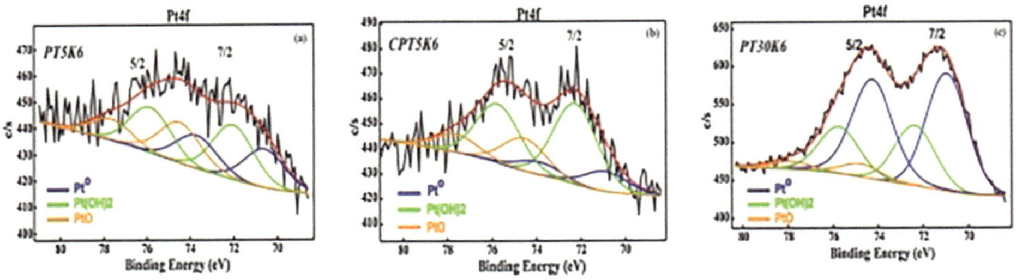

Figure 2.
XPS spectra Pt-modified KIT-6 mesoporous silica and Pt species atomic percent (with permission from Ref. [16]).

Another transition metal that is present as active component in OMSs supports was vanadium. V-MCM-41 has received many applications in oxidation reactions [3, 30–32]. For ordered mesoporous V-MCM-41 materials synthesized by DHT method [3, 25, 31], vanadium occurs mainly as isolated tetrahedrally coordinated V^{5+} species incorporated in the pore wall or anchored to the pore wall. UV–Vis spectra reveal that all the samples prepared with low V contents present well-dispersed V species in the silica network as V^{5+} species. At high surface vanadium coverage, the species are substantially polymerized. The oxidation of V^{4+} species in the precursor has also been observed. The change in the UV–Vis spectra after calcination was due to the modification in the oxidation state of vanadium (V^{5+}) from the isolated tetrahedral coordination to its distorted octahedral coordination by coming into contact with the water molecules in the atmosphere [31]. Shylesh et al. [32] reported that UV–Vis spectra of VMCM-41-DHT materials showed vanadium incorporated into the framework positions for VMCM-41 samples, while the greater percentage of active species resides on the surface of VMCM-41, enhancing the formation of higher coordinated vanadium species after calcination. Treating MCM-41 with an aqueous or alcoholic solution of vanadyl acetylacetonate can lead either to a grafting of vanadium entities on the silica surface or to an ion exchange between surfactant molecules and vanadium cations in solution. The UV–Vis spectra of the samples prepared in water or alcohol with low V contents ($V/Si < 0.1$) showed essentially two absorption bands at 275 and 345 nm. The first was assigned to V^{5+} species inside the silica walls, whereas those corresponding to the band at 345 nm located on the surface of the mesopores. The presence of internal sites is due to the reorganization of the hexagonal tubular structure of MCM-41 upon hydrothermal treatment, during which vanadium species are allowed to penetrate the silica walls. Thus, vanadium species in the samples obtained by impregnation are dispersed on the wall surface while in the samples obtained by direct synthesis they were fixed in the mesoporous framework. Therefore, the surface vanadium species supported on silica are well known to possess an isolated and distorted VO_4 structure with a single V=O terminal bond and three V–O–Si bridging bonds anchored on silica support. The distorted V^{5+} species with the bridging V–O–Si could be found in different silica environments. The formation of vanadium oxide nanodomains has also been evidenced by ESR spectroscopy. A quantitative measurement of the ESR signal intensity shows that the corresponding V^{4+} species represent only 0.1% of the total V species. The majority of the species are V^{5+}. The others species are VO^{2+} which are very well dispersed and isolated inside the pore channels of MCM materials. Raman and UV–Vis spectroscopic characterization of V-MCM-41 materials were used [33] to obtain more definitive information about the possible presence of XRD-amorphous crystalline V_2O_5 nanoparticles and surface VO*x* species for samples possessing with higher vanadium content (up to 5.3 wt. %V). DR UV–Vis spectra of Me-MCM-41 (Me = Ti, V, Cr) samples obtained by direct synthesis [34] sustained the framework incorporation of Ti^{4+} ions in the inorganic silica matrix with tetrahedral or octahedral coordination, vanadium (V^{5+}) isolated species with tetrahedral environments and monochromates, with minor amounts of dichromates as well as polychromate species, respectively. The calcination treatments had changed all the Cr(III) ions to Cr(VI). A large part of these species resided on the surface of silica mesoporous support. These results indicated that the major species formed on the Cr-MCM-41 sample were monochromates, with minor amounts of dichromates as well as polychromate species. The ESR spectrum of as-synthesized chromium-containing mesoporous silica indicated, for a large part of chromium, the presence of trivalent chromium (Cr^{3+}) in octahedral coordination.

The surface vanadium species supported on silica were well known to possess an isolated and distorted VO_4 structure with a single V=O terminal bond and three V–O–Si bridging bonds anchored on silica support. The distorted V^{5+} species with the bridging V–O–Si can be found in different silica environments. Similar studies on the surface Nb^{5+} species present in Nb–MCM-41 have revealed that the Nb atoms were predominately isolated NbO_4 species under dehydration conditions, and surface polymeric niobium species and/or bulk Nb_2O_5 are formed at high niobium loading on silica [35]. Raman spectra of Ta-MCM-41 mesoporous materials indicated [6] that the incorporation of Ta atom into the MCM-41 structure forms a distorted and isolated [TaO_4] surrounded by the [SiO_4] tetrahedrons with the presence of Ta–O–Si bridging bonds, and three types of tantalum oxide species: an isolated TaO_4 species in the MCM-41 framework, an isolated surface TaO_4 species on the MCM-41 surface, and bulk Ta_2O_5, can be present individually or coexist on the Ta–MCM-41 catalysts, and it's relative intensity was dependent on the Ta concentration.

The idea of the MCM-41 impregnation with vanadium and antimony sources was to locate Sb-V-Ox species on the high surface area of mesoporous materials with various compositions [36]. Vanadium species in the prepared samples have been estimated by UV–Vis and ESR spectroscopic study. All mesoporous matrices modified with vanadium and antimony gave rise to well-resolve signals in the hyperfine structure of ESR spectra characteristic for isolated VO^{2+} species. Such a structure was not registered in the case of V/SiO_2 sample suggesting that mesoporous support was important for the isolation of oxovanadium species. Tetrahedrally coordinated vanadium (IV) species were deduced from UV–Vis spectra on all prepared samples. They were the only registered species on SbV/NbMCM-41 and SbV/AlMCM-41, whereas on SbV/MCM-41 and SbV/SiO_2, octahedral ones were also present besides them. Octahedral coordination dominated in bulk SbVOx.

There are many studies on cobalt incorporation in mesoporous silica supports. The information about the nature, the co-ordination, and the location of the metal species for cobalt- and cobalt-vanadium-modified MCM-41 materials obtained by direct synthesis were obtained by TPR, DR-UV–Vis, and XPS analysis [25]. These methods indicated different localization of the cations in extra-framework positions or in the framework of MCM-41molecular sieves. DR-UV–Vis spectra from **Figure 3A** show two different types of V^{5+} species. The first one was assigned to isolated tetrahedrally coordinated V^{5+} species and the second originates from polymeric tetrahedral V^{5+} species grafted on the walls. According to these results, H_2-TPR measurements (**Figure 3B**) suggested that the vanadium interaction with MCM-41 was predominant in VCoMCM-41 samples and pointed to the presence of monomeric or low oligomeric dispersed tetrahedral vanadium species obtained by direct synthesis and the formation of less reducible "polymeric-like" vanadium species by postsynthesis. In the low-loaded cobalt catalysts, Co^{2+} in tetrahedral position was observed. The increase in metal content led to the appearance of Co^{3+} in Oh symmetry. In both cases, the cobalt ions were placed outside of the silica framework. In the bimetallic samples, vanadium was incorporated inside the framework of the molecular sieves and on the channel walls. V^{5+} was in tetrahedral symmetry. In the bimetallic samples, cobalt was presented as Co^{2+} in Td symmetry. When Co and V were introduced together in the starting gel, a lower quantity of vanadium was incorporated into the mesoporous sieve. At a low vanadium concentration, the essential part of cobalt gives rise to the cobalt silicate phase. The latter was reduced at higher temperature. The rest of cobalt forms CoO particles interacting weakly with the siliceous framework reduced at lower temperature. The peak at 710 K for VCo3 sample was most likely the composite one from the reduction of both cobalt and vanadium species.

Impregnation of MCM-41 and SBA-15 materials using aqueous solutions of cobalt nitrate has a significantly different impact on their ordered mesoporous structures. Thus, aqueous impregnation of MCM-41 followed in the surface area and pore volume. By comparison, SBA-15 mesoporous structure remained almost intact after the introduction of significant amounts of cobalt (up to 20%). The different behavior of these two mesoporous silicas was principally attributed to the different pore wall thickness in MCM-41 and SBA-15. Cobalt oxide-modified SBA-15,

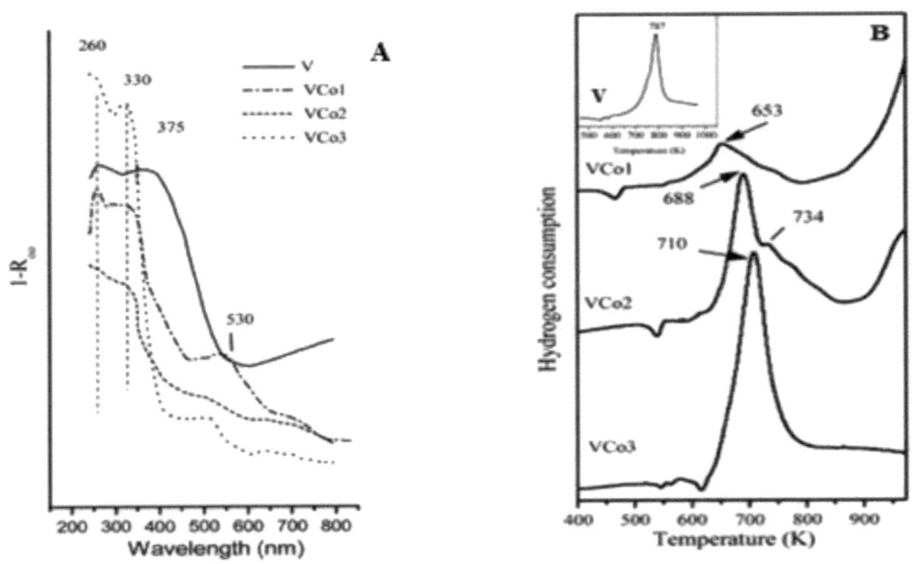

Figure 3.
DR-UV–Vis spectra (A) TPR profiles (B) of VCo-MCM-41 and V-MCM-41 samples (with permission from Ref. [25]).

KIT-5, and KIT-6 mesoporous silicas with different pore size/pore entrances have been synthesized by a conventional wet impregnation method using cobalt nitrate as the precursor. UV–Vis spectra indicated the formation of Co_3O_4 particles with different degree of dispersion, which are in different interaction with the support [2]. XPS spectra showed the variation of surface Co dispersion with aging temperature that facilitated cobalt species migration and agglomeration through the larger pores of the silica matrix. The effect of the pore size was less pronounced for the SBA-15 materials, where the straight cylindrical pores with 2-D arrangement probably leads to homogeneous distribution of the loaded cobalt oxide particles along the pore surface. For the 3D structures with interpenetrated cylindrical mesopores (KIT-6) or cage-like mesopores (KIT-5), the formation of homogeneously dispersed spinel Co_3O_4 species seems to be facilitated in mesoporous silicas with pores larger than 6 nm. TPR-DTG results evidenced the co-existence of three types of Co_3O_4 particles for all cobalt-modified SBA-15, KIT-5, and KIT-6 materials. The first type was easily reducible, relatively larger species, loosely interacting with the support; the second type represented hardly reducible and well-dispersed fraction in a moderate interaction with the silica, and the third type was very finely dispersed, strongly interacted with the support, which could not be reduced up to 873 K. The pronounced differences were observed for Co/KIT-6 materials [2]. Therefore, Co/KIT-6 samples presented a significant portion of crystalline species that weakly interacted with the support. For the KIT-6 aging to higher temperature, the presence of more inhomogeneously dispersed cobalt oxide particles, which were not completely reduced to metallic cobalt in the temperature interval 500–750 K, was revealed. For these materials, TPR-DTG analysis in correlation with the FTIR measurements supported formation of spinel-type Co_3O_4 species in the case of silicas with larger mesopores. The incorporation of Co and/or Fe in HSM and SBA-15 was evidenced by association of XRD, TEM, and TPR techniques [37]. Low-angle XRD patterns indicated that the mesoporous supports remained unchanged after metal impregnation and calcination steps; hence, still retained the ordered structure.

Iron is an interesting metal regarding its properties and applications in catalytic oxidation. In case of its immobilization in mesoporous silica by direct synthesis and hydrothermal treatment [4, 32, 38, 39], most of iron species exist in the tetrahedral coordination located in the support framework. The results obtained from DRUV–Vis, ESR, and XANES showed iron in a mixed environment, indicating that some iron was tetrahedrally coordinated, being sited in the framework, and some iron is present as an extra-framework atom, being octahedrally coordinated [38]. XANES results also suggested that copper was present in the Cu–Al-MCM-41 samples both in the framework and in the extra-framework sites as hydroxide and oxide, respectively. ESR spectra of hierarchical silica structures confirmed the presence of Fe^{3+} ions with tetrahedral coordination both in framework and extra-framework support. For samples with higher loading, the presence of interstitial oxide phase and iron oxide clusters was displayed [28]. Similar results were obtained for copper. XANES results also suggested that copper is present in the Cu–Al-MCM-41 samples both in framework and extra-framework sites as hydroxide and oxide, respectively. The presence of Al^{3+}-sites on the surface of the support provides considerably better dispersion of copper [26, 40]. When comparing ZnAl-MCM-41 with FeAl-MCM-41 samples, the interaction between the metal and the framework atoms (Zn...Si) was different. 27Al-MASNMR results have indicated that zinc is not substituted for aluminum, which means taking the EXAFS results into account that zinc (II) substitutes for silicon (IV) in the framework. The presence of several, and probably different, silicon-sites in the mesoporous framework explains the higher disorder in Zn–Al-MCM-41 compared to Fe–Al-MCM-41, in which iron substitutes for aluminum.

The incorporation of Ce, another trivalent metal, within the MCM-41 was favored by the greater flexibility of the silica network. However, the size incompatibility between Ce^{3+} and Si^{4+} ions led to longer Si\O\bonds and caused the strain bond angle in the substituted silica network. Also, the incorporation of Ce induced a drastic reduction in mesopore ordination. These results were probably due to partial substitution of the structural Si^{4+} for the Ce^{3+} ion, resulting a substantial change in the textural properties of the hexagonal structure of MCM-41 [41]. The Si/Ce molar ratio is a key factor influencing the textural properties and structural regularity of CeMCM-41 mesoporous molecular sieves. As well, XRD, UV–Vis, and XPS spectra evidenced the presence of cerium species as tetra-coordinated Ce^{4+}/Ce^{3+} and the formation of CeO_2. This was in accord with results obtained on cerium incorporated in SBA-15 [42] and KIT-6 mesoporous silica [16]. The effect of pH on SBA-15 ordered hexagonal structure and incorporation of Ce species was evidenced [42]. For the samples synthesized at pH = 10.0, the position of Ce species was evidenced as deposits only on the surface of SBA-15.

High metal dispersion and incorporation of Ni in MCM-41 framework was evidenced for lower metal loading. Typical XRD diffractograms for ordered hexagonal mesoporous structure, obtained at small angle, evidenced decreasing of structural regularity with metal loading. Considering UV–Vis of NiO as reference, a distorted tetrahedral environment was observed for the most of Ni species in these MCM-41 materials. The effect of Zr^{4+} on Ni^{2+} local symmetry and the presence of distorted tetrahedral Ti species were evidenced by UV–Vis for Ni-ZrMCM-41 and Ni-TiMCM-41 bimetallic samples. Thus, a small shoulder at around 293 nm was assigned to penta- or octahedral coordinated Ti species, resulting from the interaction of Ti species with Ni species [27]. For Ni–MnMCM-41 sample, it was assumed that both tetrahedral and octahedral Mn^{3+} species co-exist. Mn^{3+} was evidenced both in tetrahedral and octahedral coordination. The results obtained for bimetal samples were compared with them with single metal. Such for Mn-MCM-41 sample, XRD patterns showed the absence of the diffraction peaks of the MnOx species suggesting that a strong interaction between MnOx and silica matrix exists because most of the Mn^{3+} or Mn^{2+} cations were either incorporated into the silica framework or highly dispersed on the silica walls. The TPR results on Mn-MCM-41 samples [43] indicated the coexistence of different manganese species. In the samples of different pore dimensions and manganese loadings prepared by impregnation, the nature of the species, identified as well dispersed, strongly interacting with silica surface was similar. In the case of samples prepared by the hydrothermal method, the effect of pore dimensions was more complex. Narrow pores of silica materials caused the formation of small species strongly interacting with silica surface or incorporated into the framework. An increase in Mn loading and pore diameter favored formation of larger particles weakly interacting with silica support. It was observed that the presence of small oxide species of the size partially controlled by pore dimension or preparation method, and simultaneously not strongly interacting with silica support.

The incorporation of larger species into the silica framework was hindered and the formation of extra-framework oxide species was favored. Regarding the incorporation of tungsten species into the MCM-41 framework, there is a critical value for the Si/W ratio of about 30. In the case of smaller Si/W ratio, the formation of extra-framework tungsten oxide species was observed [44]. Variation of the Si/W ratio and the synthesis method has led to various species of W immobilized on HMS silica [45]. Thus, through Raman spectroscopy, isolated $[WO_4]^{2-}$ or low condensed oligomeric framework species were displayed. Tin is another metal with redox properties and large size which forms SnO_2 clusters distributed on the external pore structure. SnO_2 agglomerates were highlighted in the channels or on the external surface, which blocked the pores partially, thereby reducing the surface area. By adjusting the $nH_2O/nHCl$ molar ratio, Sn was incorporated into the lattice of SBA-15 at a low Sn concentration [46]. The Sn^{4+} ions exhibited both tetrahedral and octahedral coordination depending upon the location of these ions either on the walls of the silica or in the corona region of the structure, respectively. The existence of isolated oxide species that have degraded the ordered structure of the silica support and especially the formation of the oxide agglomerations in the pores or on the external surface has been highlighted for other metals with large diameter (Ru, La) but very active in oxidation reactions [24, 47–50].

The imobilization of active metals in the specific locations of ordered mesoporous silicas by direct synthesis routes with the help of organic groups of surfactants brought a new aspect of creating metal-functionalized OMSs [51]. Although the strong interactions between active metals and support were obtained, the controllable morphology and structure of OMSs synthesized by these direct synthesis routes have not been well developed. The synergistic effect between loccation, its dispersion, and mesoporous ordered silica structure on the metal electronic properties and catalytic needed development of the advanced characterization techniques.

3. Catalytic oxidation of organic compounds

The introduction of the metal cations in the mesoporous silica generated both acid and redox centers depending on their charge and their chemical properties. In oxidation reactions, these properties determine both the activity and the selectivity of the catalyst. To introduce the redox active sites in the OMSs, various transition metals have been chosen. The supports like M41S, SBA-n, and KIT-n families modified by incorporation of one, two, or more transitional metals such as Ti, V, Cr, Fe, Co, Ni, Mn, Cu, La, Ru, Ni–Ru, Cr–Ni, V–Cu, and V–Co created materials with new redox and acidic properties. The introduction of active transition metal into the framework of molecular sieves creates isolated metal sites and these centers are believed for their exceptional catalytic activity. Their catalytic properties were influenced by localization and surroundings of the metal ions. The high dispersion of metals on a support with high surface area, large pore diameter, and uniform pore size distribution determined the formation of new active centers with redox properties different from those of the oxide in the agglomerated form. This explained the increased interest in them and their applications as catalysts.

Table 1 shows a wide range of metals incorporating in mesoporous silica supports with catalytic applications in liquid phase oxidation of organic compounds, with H_2O_2 or *tert*-butyl hydroperoxide, and gas phase with O_2 from air. Various publications have shown the effects of metals and their associations with silica support and other metal on catalytic activity and selectivity. Thus, a high variety of transition-metals incorporated in mesoporous silica showed interesting catalytic properties in oxidation of organic compounds. Among them, vanadium and titanium were mostly used both single as well as associated with other metals. Vanadium-containing mesoporous materials are found to be active in liquid-phase oxidation reactions as oxidation of cyclohexane to cyclohexanone and cyclohexanol [31], oxidation of aromatic hydrocarbons and alcohols [3] using H_2O_2 as oxidant. V-MCM-41 catalysts exhibited low activity in the oxidation of alcohols but higher activity and selectivity in oxidation of cyclohexene and aromatic hydrocarbons. This suggested the association of vanadium with another metal more suitable for other oxidation reactions [21, 24]. V-TiMCM-41, V-CoMCM-41 catalysts were used in oxidation of aromatic hydrocarbons and alcohols [21, 24]. In these reactions, FeMCM-41, CoMCM-41, NiMCM-41 [3], NbMCM-41, Nb-TiMCM-41, Co-(Nb,

Catalyst	Reaction
VMCM-41	Oxidation of cyclohexane to cyclohexanone and cyclohexanol [31], oxidative dehydrogenation of propane (O_2 and N_2O as oxidant) [33]
V-MoMCM-41, Cu-FeMCM-41	Oxidation of o-xylene to phthalic anhydride in air [1], oxidation of adamantane with H_2O_2 [39]
VMCM-41, FeMCM-41, CoMCM-41, NiMCM-41	Oxidation of aromatic hydrocarbons and alcohols with H_2O_2 [3, 4, 30, 32]
VMCM-41, NbMCM-41, V-TiMCM-41, Nb-TiMCM-41, Co-(V,Nb,La)MCM-41, V-CoMCM-41	Oxidation of aromatic hydrocarbons and alcohols [21, 24]
MeMCM-41 (Me = V, Fe, Co, Ni), WHMS	Oxidation of styrene and benzene with H_2O_2 [4, 45]
Ru-(Cr, Ni, or Cu) MCM-41,La-(Co or Mn) MCM-41	Oxidation of aromatic hydrocarbons with H_2O_2 [49, 50]
MMCM-41 (M = Ti, V, Cr)	Oxidation of ethyl benzene and diphenyl methane with H_2O_2 [34]
TaMCM-41	Oxidative dehydrogenation of propane and oxidation of methanol [6]
CeMCM-41, CeKIT-6	Hydroxylation of 1-naphthol with H_2O_2 or *tert*-butyl hydroperoxide [41], oxidation of cyclohexene [52]
M (M = Al, Zr, W, B, or P) MCM-41	Oxidation of ethane [40]
Pt-SBA-15, Pt-NH2SBA-15, Pt-CoSBA-15, Pt-Co-NH2SBA-15	3-Butene-1-ol, *cis*-2-butene-1, 4-diol, cyclohexene oxidation with H_2O_2 [11, 30]
PtSBA-15/PtSiO_2	Oxidation of toluene with O_2 [12]
CoSBA-15, CoKIT-5, CoKIT-6	Ethyl acetate total oxidation [2]
LaKIT-6, La-BKIT-6	Oxidation of styrene with H_2O_2 [15]
TiKIT-6, Pt-TiKIT-6, Ce-TiKIT-6, Pt-Ce-TiKIT-6,	Methane in air [16]
M/KIT-6 (M = Mn, Cu, Fe, Cr, Sn)	Catalytic combustion of chlorobenzene [18]
CuImph (Imph = bis(4-imidazolyl methyl) benzylamine) on MSNs	Oxidation of toluene in air [19]
([Cu(acac)(phen) (H_2O)](ClO_4), [Cu(acac)(Me_2bipy)](ClO_4)) on HSM or NH_2HMS	Oxidation of aromatic compounds (anisole, phenol) with H_2O_2 [20]

Table 1.
Catalytic applications of modified mesoporous silica ordered networks with transitional metal.

La)MCM-41 [21, 24, 50] and Ru-(Cr, Ni, Cu)MCM-41, La-(Co, Mn)MCM-41 [49], and WHMS were also used as catalysts [45]. It was interesting to note that samples which are active in the oxidation of styrene to benzaldehyde would be less active in the oxidation of benzene to phenol and vice versa, suggesting that active centers for oxidation of styrene might probably be different to those for oxidation of benzene. Reaction data showed that the oxidation activity is higher when H_2O_2 is used as an oxidant, acetonitrile as solvent, and V-MCM-41 as catalyst. However, the selectivity toward the desired keto derivatives (ethyl benzene to acetophenone and diphenyl methane to benzophenone) follows the order, Ti-MCM-41 > V-MCM-41 > Cr-MCM-41 [34]. The vanadium content in catalysts was evidenced as a key factor in the oxidation of styrene and the conversion increases with metal loading. However, for hydroxylation of benzene, catalysts with an ordered mesostructure presented higher catalytic activity and the V content only serves as secondary role.

It can be concluded that high vanadium content favored styrene conversion and higher ordering of mesostructure led to high hydroxylation of benzene. The higher activity of vanadium-incorporated MCM-41 compared to vanadium-grafted MCM-41 may be due to the presence of active isolated tetrahedral-coordinated vanadium ions in the framework positions. The lower activity was a result of the V–O–V bond formed for vanadium-grafted MCM-41. The difference in the selectivity of as-synthesized and calcined vanadium-grafted MCM-41 showed that apart from the active redox sites, the nature of hydrophilic–hydrophobic interactions still play an important role in selective oxidation reactions. **Figure 4** depicts the variation in styrene conversion and reaction rate as a function of reaction time. The conversion and reaction rate in oxidation of styrene were modified by the introduction mode of the H_2O_2 into reaction medium in order to find the optimal reaction conditions.

The higher conversion and reaction rates were obtained in the first period of reaction when H_2O_2 was introduced after adsorption the step (samples VTi-2 and NbTi-2). These results sustained the effect of the adsorption step on oxidation reaction confirmed, over time, by photocatalytic reactions, other oxidation processes in which both Ti and V catalysts or other transition metals with redox properties were used immobilized on mesoporous silica [7–10, 28]. An induction period was needed (VTi-1, NbTi-1catalyst) in oxidation of aromatic hydrocarbons when the oxidant, reagent, catalyst, and solvent were mixed together at the beginning of the reaction and effect of the second metal was significant. The effect of the second metal on the properties of catalysts was evidenced for others materials. Firstly, the introduction of V and La reduced sharply the conversion of styrene.

The increasing of Co/V molar ratio (decreasing V amount) led to increasing of the catalytic activity. The highest conversion was obtained for Co/V molar ratio of 3.0. On the other hand, the effect of the second metal on the reactivity was different for styrene and benzene. The introduction of V or Nb into CoMCM-41 molecular sieves decreased or increased the conversion of styrene. The effect of La on CoMCM-41 was quite surprising since the activity for oxidation of styrene and benzene decreased, implying the inhibitory effect of La for oxidation of aromatics [21, 24, 50]. The main products of reaction were benzaldehyde, for oxidation of styrene, and phenol for oxidation of benzene. After the first utilization, the separated and dried catalysts were reused. The catalytic activity of these reused catalysts increased in the second cycle of reaction and decreased after the third. It was observed that the selectivity decreased gradually with reaction time and in the second cycle of reaction. Some

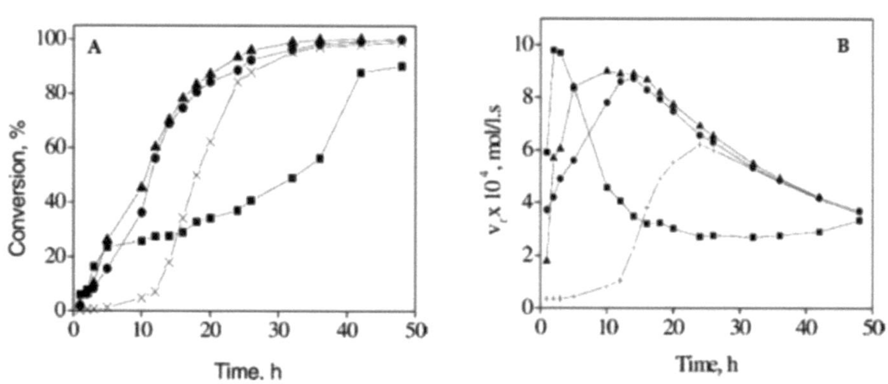

Figure 4.
Variation of the styrene conversion (A) and of the reaction rate (B) as a function of reaction time (h). VTi-1 and NbTi-1: the defined amount of H_2O_2 was introduced at the beginning of the reaction and VTi-2 and NbTi-2: the defined amount of H_2O_2 was divided in different portions and introduced step by step ((×) VTi-1, (●) VTi-2, (■) NbTi-1, (▲) NbTi-2) (with permission from Ref. [21]).

products of polycondensation resulted by polymerization of formaldehyde and/or benzaldehyde were detected after long time of reaction, a possible reason for the decrease in selectivity with reaction time. The characterization of the used and reused materials confirmed, in many cases, the good stability of bimetallic molecular sieves. TEM images of the used catalysts showed a well-ordered structure with a hexagonal arrangement of channels, indicating no effect of the reaction on the support structure after first and second reaction [24]. The IR study of adsorbed phase in discharged catalysts after first cycle reaction of styrene and benzene upon desorption at a series of temperatures (293, 373, 623, and 723 K) was made [50]. The presence of different aromatic species such as benzaldehyde and styrene glycol was observed. These species adsorbed very strongly in the catalyst since a complete desorption of these species can be made only after desorption at 723 K. The strong adsorption of these species was confirmed by thermal analysis.

The association of trivalent cations Ru or La with other transition metals modifies the activity and selectivity of the monometallic molecular sieves [24, 49]. The results evidenced that RuMCM-41 and LaMCM-41 presented good conversion in oxidation of styrene but very low activity for benzene hydroxylation. While all the bimetallic catalysts, except Ru-NiMCM-41 and La-CoMCM-41, have a higher activity and efficiency of the H_2O_2 (H_2O_2 quantity used for oxidation/H_2O_2 quantity transformed) in the benzene hydroxylation. In the styrene oxidation, only RuCr-MCM-41 gives a good conversion. Generally, excepting LaCo-MCM-41, the conversion of catalysts in oxidation of benzene was higher than that in oxidation of styrene. This behavior was specific for these bimetallic samples since all the monometallic modified MCM-41 catalysts by incorporation of the same transition metals have a higher activity in oxidation of styrene and lower conversion in hydroxylation of benzene. LaCo-MCM-41 catalyst has a very low activity and RuCr-MCM-41 catalyst has a high activity both in the styrene and benzene oxidation. Under all investigated experimental conditions for oxidation of styrene, the main reaction products detected by GC analysis were epoxy ethyl benzene (styrene oxide), phenyl ethanediol (styrene glycol), and benzaldehyde.

More transition ions were incorporated into MCM-41, HMS, SBA-15 materials, and tested in liquid-phase oxidation reactions (**Table 1**). The best activity in oxidation of benzene was obtained for Ti-MCM-41 and in oxidation of styrene for Cr-MCM-41 and CrNi-MCM-41 samples. Their activity decreases with increasing of number of 3d electrons of the metal ions. After immobilization and interaction with support and another metal, it was highlighted that increasing of their oxidation state leads to the growth of their activity. Ti, V, Cr, and Mn ions were more active. After immobilization and interaction with support and another metal, an increase of their oxidation state has been observed. The obtained redox molecular sieves by direct synthesis incorporation of tungsten into hexagonal mesoporous silica (HMS) were tested in oxidation of styrene with hydrogen peroxide. The influence of the synthesis parameters, such as, chemical composition of the gels, surfactant, precursors, and time of the hydrothermal treatment, on the structure, morphology, nature of metal species, and catalytic properties has been examined. The best results were obtained for the catalysts synthesized by oxo-polyoxo mode: ($[WO(O_2)_2(H_2O)_2]$/H_2O/H_3O^+/surfactant/$Si(OR)_4$ in which surfactant was cetylpyridine chloride [45]. The catalysts with more ordered structure, higher surface area, and $[WO_4]^{2-}$ species strongly bounded or high dispersed on silica have a higher activity. Conversion and selectivity to benzaldehyde decreased with Si/W molar ratio (**Figure 5**). The higest conversion of styrene to benzaldehyde has been evidenced the sample with the highest amount of isolated W sites and higher surface area.

In addition to applications in the oxidation reactions [11, 24, 31], cobalt-based catalysts were considered as a suitable alternative to the high cost of catalysts based

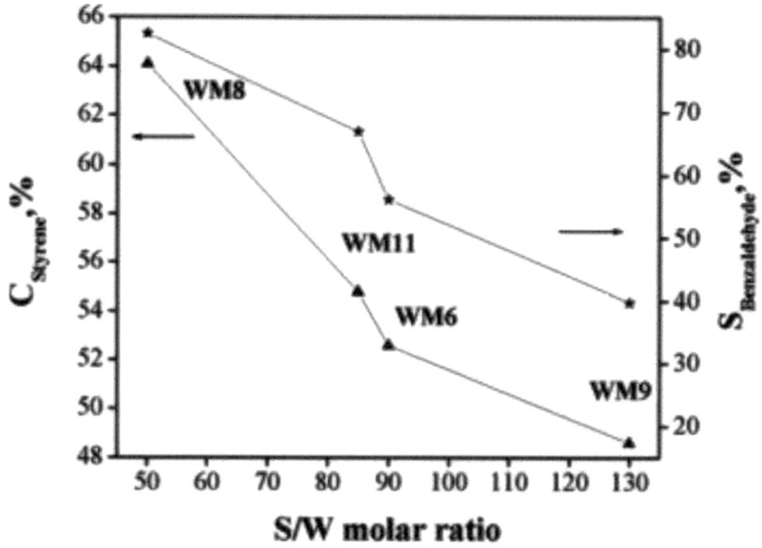

Figure 5.
Variation of conversion and selectivity with Si/W molar ratio (with permission from Ref. [45]).

on noble metals. The catalysts obtained by immobilization of Co on mesoporous silica were used for degradation of toxic compounds from the exhausted automobile and industrial emissions via combustion. Cobalt-modified SBA-15, KIT-5, and KIT-6 mesoporous silicas with different pore sizes/pore entrances have been synthesized and their catalytic activities in total oxidation of ethyl acetate were evaluated [2]. The optimal size of Co_3O_4 particles and their homogeneous distribution in the porous support are of primary importance for the catalytic activity. Silicas with larger pores facilitate the mass transfer during the sample preparation procedure and lead to more homogeneous distribution of cobalt oxide particles inside the pore system. The Co_3O_4 particle growth is facilitated by 3D structures with interpenetrating-(KIT-6) or cage-like (KIT-5) pores and much less promoted by 2D arranged straight pores of SBA-15 support.

Furthermore, more open pore structures could facilitate the catalytic process due to enhanced mass transfer, as the catalytic activity of CoKIT-6 materials is generally lower than that of CoSBA-15 and CoKIT-5.

Cerium-containing materials have been used as catalysts for selective oxidation of organic compounds. The liquid-phase oxidation of cyclohexane was carried out over Ce-KIT-6 with various Si/Ce molecular sieves at temperature between 70 and 90°C. Although the catalyst contains both Ce^{3+} and Ce^{4+}, as evidenced from DR-UV analysis, the main active sites that activate H_2O_2 were considered to be Ce^{4+} [52]. Ce^{4+} was suggested to activate hydrogen peroxide by coordination to oxidize cyclohexane, and cyclohexanol was found to be the major product (74%). Cyclohexyl acetate, as the major secondary product, was considered result of Bronsted acidity generated by Ce^{3+}. Cerium and titanium oxides, immobilized on KIT-6 silica, were used as supports for Pt. The supported Pt nanoparticles on mesoporous silica possess the ability to strongly dissociate toluene to benzene and hydrocarbon fragments (CHx) [12]. The metal–support interaction was evaluated in terms of TiO_2 loading and ceria effect on titanium oxide under condition of their dispersion on silica [16]. The highest activity was obtained in oxidation of CH_4 for all the catalyst samples containing Pt respectively for PT5K6 sample with lower dispersion of Pt and higher diameter of particles (**Figure 6**). Two opposite effects determined in this case a very small decrease in the CH_4 conversion degree

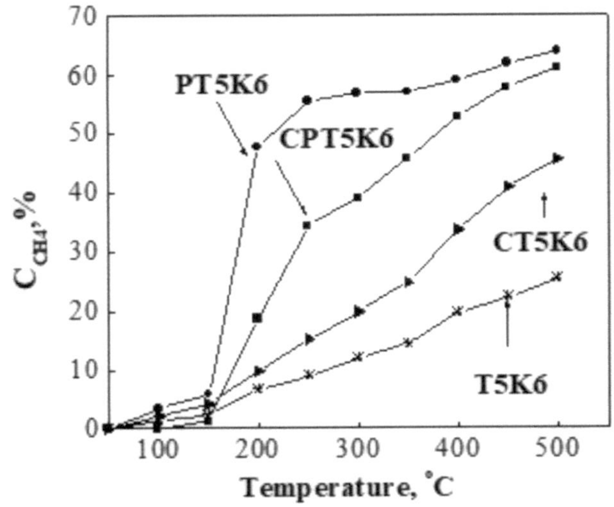

Figure 6.
CH_4 conversion and its variation with temperature for catalysts with different composition: Supported on T5K6 (a), CPTnK6 (B), and TnK6 (C) (with permission from Ref. [16]).

compared to PT5K6 catalyst. The higher concentration of Pt^{2+} on the catalyst surface (CPT4K6 samples) and absence of noble metal (CT5K6, T5K6 samples) decreased the catalytic activity and a slow growth of conversion with temperature was observed. A good correlation was obtained between catalytic activity and reducibility of the catalysts for samples with Ti and Ce. The higher conversion degree of CH_4 was obtained for PtTi-KIT-6 samples between 250 and 400°C. KIT-6 was an interesting support for other transition metals such as Mn, Cu, Fe, Cr, Sn, Ln, and the obtained catalysts were used in oxidative degradation of various organic pollutants from air (toluene, chlorobenzene) [18].

4. Incorporated metal active sites with redox properties

It can be considered that the interest on catalytic performances obtained by incorporating titanium into the zeolite (TS-1 catalyst) and the advantages of silica mesoporous molecular sieves have generated research on the metal-functionalized ordered mesoporous silicas. These catalysts have attracted much attention in the past few years for selective oxidation and oxidative photodegradation of large organic molecules. This explains the particular interest in the synthesis and use of catalysts based on titanium immobilized on various mesoporous silica substrates. The active center of TS-1 catalyst in the selective oxidation reactions was considered the titanium-isolated sites. A systematic study on the function of the different titanium species in TS-1 [53] concluded that the framework Ti species in TS-1 were the active sites for propylene epoxidation, while the nonframework Ti species were responsible for the further conversion of propylene oxide to propylene glycol and methoxy-2-propanol. Similar to TS-1, the remarkably catalytic properties of the metal-functionalized ordered mesoporous silicas result from the isolation of the metal centers in the framework. The high surface area of mesoporous sieves and the presence of ordered arrays of mesopores provide new opportunities for transition metal incorporation. It is possible to obtain metal heteroatoms as framework tetrahedral T atoms, bound in defect sites of the framework, anchored to the surface, extra-framework counter ions or extra-framework oxides.

The isolated heterogeneous single-site catalysts have been attracted great interest in diverse catalytic reactions because of their uniform and distinct geometric and electronic structure [54]. They have great potential to integrate the distinct advantages of homogeneous and heterogeneous catalysts into a single counterpart. The imobilization of active metals in the specific locations of ordered mesoporous silicas by direct synthesis routes with the help of organic groups of surfactants opened a new path in creating of metal-functionalized OMSs. Although the strong interactions between active metals and support were obtained, the controllable morphology and structure of OMSs synthesized by these direct synthesis routes have not been well developed. It is still difficult to understand the relationship between the structural morphology of OMSs and the electronic structure of active species, thus the development of advanced characterization techniques was necesary. Similar to other supported single sites [54], it can be considered that their catalytic activity is due to the following which allows: the ordered arrangement of silica support mesoporous structure; the high dispersion of metal sites, located in framework and extra-framework support, or covalently bounded on the pore surface functionalized with different ligands; the better contact of single site with the reactants thus generating catalysts with high activity and excellent selectivity; the ordered porosity can provide a special environment for the substrate interaction with the catalytic active sites which can further enhance the activity and selectivity.

Considering Ti (IV) species to be the active sites in TiMCM-41 catalyst in the oxidation of cyclohexene to cyclohexene epoxide, a mechanism in three steps was proposed for surface reaction [55]. The main reaction step involved the reaction of Ti single sites with H_2O_2 to form titanium hydroperoxide, which further reacts with cyclohexene molecules to form cyclohexene epoxide in a concerted manner. It was evidenced that titanium-isolated species from silica framework or surface were the active species. According to titanium location, two Ti-peroxide (η^1 and η^2) intermediates wherein Ti binds 1, respectively both oxygen atoms of the peroxide (**Figure 7**), have been identified.

The stability and reactivity of Ti-peroxide intermediates was affected by solvent coordination. The best results were obtained in acetonitrile. The extra-framework Ti species were amorphous or crystalline TiO_2. These species had a negative effect on the yield of propane oxide. The amorphous Ti species were more acidic and mainly responsible for the further conversion of propane oxide.

The mechanisms proposed for oxidation of organic compounds with H_2O_2 on vanadium-modified mesoporous silica supports proposed also the formation of V-peroxide intermediates (η^1 and η^2). These results were also generalized for the activity of other immobilized metals. Thus, the specific catalytic activity (per one Cu^{2+}-active site accessible to the reactants) depended strongly on the structure of

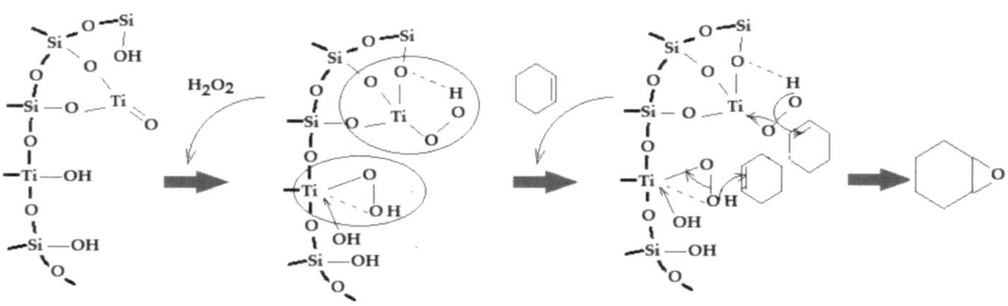

Figure 7.
The oxidation mechanism proposed for possible isolated Ti^{4+} active sites immobilized on mesoporous silica.

the localized site [40]. Isolated Cu^{2+}-sites grafted to Al-MCM-41 showed relatively high activity for the sample calcined at lower temperature. Thermal treatment at higher temperatures caused a sharp loss of specific Cu^{2+} catalytic activity of CuAl-MCM-41 as result of copper oxide agglomeration.

If the organic substrate was introduced in the first step of the reaction, it was adsorbed on the surface and reacted more rapidly with the Me-peroxide intermediate formed by the addition of H_2O_2 in the second step [21]. The presence of the organic compound on the surface prevented H_2O_2 decomposition more in the presence of extra-framework oxides that favor its decomposition. Depending of their loading and dispersion, the activity of these metal extra-framework species was similar with that of clusters or bulk oxides. For Ru- and La-incorporated MCM-41 molecular silica, the ratio of metal and oxygen ion radii ($RM(n^+)/RO(2^-) > 0.5$), an important structural parameter, showed a little possibility to be incorporated metals in the silica framework [49]. For many of these catalysts, the efficiency of the H_2O_2 (H_2O_2 quantity used for oxidation/H_2O_2 quantity transformed) was very low. These catalysts had a good conversion in oxidation of styrene but very low for benzene hydroxylation. Their association with metals with smaller diameter as Co or Mn led to a higher activity and efficiency of the H_2O_2.

The hydrophobicity of metal-functionalized ordered mesoporous silicas was improved, in many cases, by functionalization of surface with organic groups in direct (co-condensation) or postsynthetic silylation treatment. The coexistence of metal cations with other organic or inorganic species favored the distribution on the pore surface of catalytic active species, thus favoring the access of the reactants during the chemical reaction. However, in this case, the metal species formed oxide agglomerations, which led to the decreasing of the conversion. Thus, for LaKIT-6 catalyst, the coexistence of B hindered the dispersion or incorporation of La into the framework [15]. La^{3+} species in the framework of LaKIT-6 were favorable to improve the catalytic performance of LaKIT-6 for the oxidation of styrene. When H_2O_2 was used as an oxidant, LaBKIT-6 showed much lower styrene conversion than LaKIT-6 catalyst with similar product selectivity, which was ascribed to the formation on surface of La_2O_3 nanocrystals.

The controllable dispersion of metal single sites and a higher hydrophobicity were obtained by immobilization of the metal complexes on mesoporous silica supports. A number of mono- and binuclear metal complexes have been investigated as biomimetic catalysts for organic compound oxidation. New biomimetic catalysts were obtained by immobilization of $[Cu(AcAc)(Phen)(H_2O)]ClO_4$, $[Cu(AcAc)(Me_2bipy)]\ ClO_4$ complexes on HMS or NH_2-functionalized mesoporous silica [20]. The copper-substituted mesoporous silica was a good catalyst for oxidation of aromatic compounds or a very good support for biomimetic oxidation catalyst due to the possible interactions between the metallic ions of the biomimetic complex and the stabilized cuprous species in the silica framework. The immobilization on different mesoporous silica supports of (Schiff base) copper(II) or Mn(II) complexes was synthesized and applied in oxidation of alcohols in acetonitrile and H_2O_2. Comparing the effect of silica supports on catalytic activity, the higher performances of the metal complexes supported on MCM-41 were evidenced. A higher catalytic activity was obtained for Mn complexes, especially in oxidation of cyclohexene in conditions of significant lower M/L ratio and metal content. The most probable mechanism for attachment of the aminopropyl groups to the surface of mesoporous silicas was proposed through siloxane linkages (Si–O–Si) between the silicon of the aminopropylsilane group and the surface silicon atoms. It has been assumed the possibility that each aminopropylsilane silicon attaches to the internal surface of the silica via one, as well two and/or three siloxane linkages. **Figure 8** proposes the metal complex structure immobilized on SBA-15 mesoporous support.

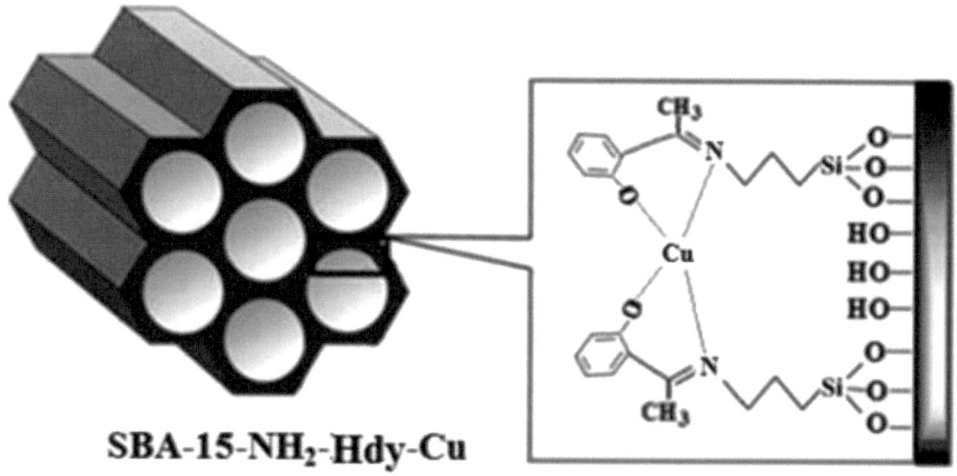

Figure 8.
Copper single site supported by complexation with Schiff-base ligands obtained from 2-hydroxyacetopheneone covalently attached to amino-functionalized SBA-15 (unpublished).

The stability and heterogeneous nature of these catalysts in the oxidation of different substrates have been investigated. The oxidation in the liquid phase with organic hydroperoxides or even H_2O_2 was often the subject of leaching phenomena and the question about the true nature of the catalytic reaction (homogeneous or heterogeneous) was a serious one.

The photocatalytic activity of supported transition metals (Ti, V, Ce, Fe, Cu, Au, Ag, Ta, Nb) on mesoporous silica supports in mono- or bimetallic photocatalysts was evaluated in degradation of various organic pollutants from water or air. Among the various mesoporous silica materials, MCM-48 and KIT-6 were several advantages in photocatalysis due to their cubic arrangement of the three-dimensional mesopores and their considerable amount of surface silanol groups. The photocatalytic properties of CeTi-MCM-48 photocatalysts were evaluated by phenol photodegradation under UV irradiation at two different wavelengths and the photocatalytic activity was correlated with the active metallic distribution, speciation, and their immobilization method [7, 8]. The incorporation of Ce in the Ti-MCM-48 framework allowed activation by UV light, generating more electrons and holes in the photocatalytic redox reactions and a better photoactivity for phenol. The presence of redox couple Ce^{4+}/Ce^{3+} exchaged the surface oxygen vacancies that could be transformed in oxygen radicals with formation of superoxide radicals. Many other photocatalysts were obtained by immobilization of the photoactive transition metals on various mesoporous silica supports, and their activity was evaluated in oxidation of organic polutants from water as dyes, functionalized 8-hydroxyquinolinate, antibiotics [9, 10, 28, 56]. In all of these, the active photoactive metallic sites were Ti^{4+}, Co^{2+}, Fe^{3+}, Al^{3+}, Zn^{2+}, Eu^{3+}, Tb^{3+}, Er^{3+}, Nd^{3+} with various dispersion on MCM-41, SBA-15 or hierarchical silica supports.

In case of the catalysts used in gas-phase reactions, dispersion, specific surface area, and reducibility was observed as the main factors influencing the catalytic activity. In combustion of chlorobenzene (CB), catalytic activity of metal-modified KIT-6 (M = Mn, Cu, Fe, Cr, Sn) mesoporous catalysts was basically in line with their redox properties, with the exception of CrKIT-6, which differs from this trend due to the aggregation of Cr species in the pore structure and surface of the KIT-6 [18]. The proposed mechanism was follows: CB was adsorbed to the catalyst by physical adsorption, then the adsorbed CB molecule dissociated on the active sites (mainly metal oxides) via nucleophilic attacks on C–Cl bond. The results obtained showed decreasing catalytic activity in the following order: MnKIT-6 > CrKIT-6 > CuKIT-6 > FeKIT-6 > Sn/KIT-6, in correlation with redox properties of the catalysts. Therefore, the best catalytic combustion activity of MnKIT-6 catalysts benefited from their large specific surface area, good Mn dispersion, better reducibility, and large amount of chemisorbed oxygen species. However, as the content of Mn species increasing, due to increased agglomeration of Mn, more MnOx clusters gave rise to pore plugging and cover adsorption sites, leading to weaker catalytic activity. Mesoporous silica KIT-6 with 3D interconnected mesopores offered a confined environment, a better dispersion of the oxide phase, and a faster diffusion of the reactants and products. The Mo/KIT-6, Fe/KIT-6, and Mo–Fe/KIT-6 were tested as potential alternatives to noble metals in the conversion of MCP [57]. Only the isolated tetrahedral Fe ions and highly dispersed small FeOx nanoclusters were responsible for the endocyclic C-C bond rupture at substituted C-C. The synergy between Fe and Mo was not observed at low temperatures and for total conversion but only for ring-opening reactions at higher temperature.

In particular, the good performance of the Ti(IV)-doped SBA-15-supported catalysts in CH4 oxidation was due to the combination of Ti(IV) structurally incorporated into the silica lattice and present as surface-dispersed TiO_2 particles. The negative effect of the Ti(IV) over the HMS-supported catalysts was related to the high acidity induced by the more homogeneous incorporation of Ti(IV) into the silica structure. The loss of catalyst activity during CH_4 oxidation reaction was not necessarily related to the sintering of the active sites but rather to the variation of the oxygen storage capacity and oxygen mobility strictly related to the support properties. For PdTi-SBA-15 catalysts, combustion of methane occurs through a redox or Mars van Krevelen mechanism was accepted [58]. Accordingly, during this reaction, PdO was locally reduced to Pd by methane, producing H_2O and CO_2, and then Pd was reoxidized by oxygen. Although the active species was considered PdO, it was widely recognized that metallic Pd plays the important role in decomposing and activating the methane molecule. Similar conclusions were obtained also for PtTiKIT-6 catalysts [16]. Therefore, the efficiency of all process was related to the redox properties of the catalyst and to the oxygen mobility. The maintenance of the activity, in spite of the significant palladium oxide sintering, was attributed to the good interaction between surface titania and silica. Such interaction favored the formation of Ti–O–Si linkages with an increase of the oxidizing potential of the Ti(IV) cations. In the presence of a support metal interaction, the increase of the PdO particle size [58], respectively PtO [16] after the long reaction is not detrimental. Another effect of metal–support interaction was observed for Ce-TiKIT-6 samples. In this case, two factors were influencing the Pt activity: dispersion and size of the platinum species and their concentration on the surface. Two opposite effects determined a decreasing in the CH_4 conversion. The higher concentration of Pt^{2+} and lower Pt^0 concentration on the catalyst surface. A good correlation was obtained between catalytic activity, and reducibility of the catalysts was also obtained for samples with Ti and Ce supported on KIT-6.

5. Conclusions

It can be concluded that OMSs containing transition metals showed extremely promising redox properties in the oxidation of organic compounds with larger molecules thanks to the mesoporous support. The applications and performances of the redox couple diversity resulted by association of transition metals on an increasing diversity of mesoporous or hierarchical ordered structures based on silica offer new research directions in this field. However, the decoration of multiple metal active species and their synergistic interactions on the surface of mesoporous silica remain to be explored in the future.

Author details

Viorica Parvulescu
Institute of Physical Chemistry "Ilie Murgulescu" of the Romanian Academy, Bucharest, Romania

*Address all correspondence to: vpirvulescu@icf.ro

IntechOpen

© 2019 The Author(s). Licensee IntechOpen. This chapter is distributed under the terms of the Creative Commons Attribution License (http://creativecommons.org/licenses/by/3.0), which permits unrestricted use, distribution, and reproduction in any medium, provided the original work is properly cited.

References

[1] Selvaraj M, Lee TG. A novel route to produce phthalic anhydride by oxidation of o-xylene with air over mesoporous V-Mo-MCM-41 molecular sieves. Microporous and Mesoporous Materials. 2005;**85**:39-51. DOI: 10.1016/j.micromeso.2005.05.046

[2] Tsoncheva T, Ivanova L, Rosenholm J, MikaLinden M. Cobalt oxide species supported on SBA-15, KIT-5 and KIT-6 mesoporous silicas for ethyl acetate total oxidation. Applied Catalysis B: Environmental. 2009;**89**:365-374. DOI: 10.1016/j.apcatb.2008.12.015

[3] Parvulescu V, Anastasescu C, Su B-L. Vanadium incorporated mesoporous silicates as catalysts for oxidation of alcohols and aromatics. Journal of Molecular Catalysis A: Chemical. 2003;**198**:249-261. DOI: 10.1016/S1381-1169(02)00694-5

[4] Parvulescu V, Su B-L. Iron, cobalt or nickel substituted MCM-41 molecular sieves for oxidation of hydrocarbons. Catalysis Today. 2001;**69**:315-322. DOI: 10.1016/S0920-5861(01)00384-4

[5] Yang Y, Lim S, Wang C, Harding D, Haller G. Multivariate correlation and prediction of the synthesis of vanadium substituted mesoporous molecular sieves. Microporous and Mesoporous Materials. 2004;**67**:245-257. DOI: 10.1016/j.micromeso.2003.11.010

[6] Jehng J-M, Tung WC, Huang C-H, Wachs IE. Structural characteristics and reactivity properties of the tantalum modified mesoporous silicalite (MCM-41) catalysts. Microporous and Mesoporous Materials. 2007;**99**:299-307. DOI: 10.1016/j.micromeso.2006.09.036

[7] Mureseanu M, Parvulescu V, Radu T, Filip M, Carja G. Mesoporous CeTiSiMCM-48 as novel photocatalyst for degradation of organic compounds. Journal of Alloys and Compounds. 2015;**648**:864-873. DOI: 10.1016/j.jallcom.2015.07.078

[8] Mureseanu M, Filip M, Somacescu S, Baran A, Carja G, Parvulescu V. Ce, Ti modified MCM-48 mesoporous photocatalysts: Effect of the synthesis route on support and metal ion properties. Applied Surface Science. 2018;**444**:235-242. DOI: 10.1016/j.apsusc.2018.03.053

[9] Kong L-L, Yan B, Li Y-J, Li Y. Photoactive metallic (Al^{3+}, Zn^{2+}, Eu^{3+}, Tb^{3+}, Er^{3+}, Nd^{3+}) mesoporous hybrid materials by functionalized 8-hydroxyquinolinate linkage covalently bonded SBA-15. Microporous and Mesoporous Materials. 2010;**135**:45-50. DOI: 10.1016/j.micromeso.2010.06.009

[10] Suraja VP, Yaakob Z, Binitha NN, Resmi MR, Silija PP. Photocatalytic degradation of dye pollutant over Ti and Co doped SBA-15: Comparison of activities under visible light. Chemical Engineering Journal. 2011;**176-177**:265-271. DOI: 10.1016/j.cej.2011.05.071

[11] Niculescu V, Aldea N, Rednic V, Parvulescu V. Platinum mesoporous silica catalysts for liquid media oxidation. Analytical Letters. 2019;**52**:5-19. DOI: 10.1080/00032719.2017.1421214

[12] Lai YT, Chen TC, Lan YK, Chen BS, You JH, Yang CM, et al. Pt/SBA-15 as a highly efficient catalyst for catalytic toluene oxidation. ACS Catalysis. 2014;**4**:3824-3836. DOI: 10.1021/cs500733j

[13] Kim TW, Ryoo R, Kruk M, Gierszal KP, Jaroniec M, Kamiya S, et al. Tailoring the pore structure of SBA-16 silica molecular sieve through the use of copolymer blends and control of synthesis temperature and time.

The Journal of Physical Chemistry. 2004;**108**:11480-11489. DOI: 10.1021/jp048582k

[14] Karthikeyan G, Pandurangan A. Post synthesis alumination of KIT-6 materials with Ia3d symmetry and their catalytic efficiency towards multicomponent synthesis of 1H-pyrazolo[1,2-]phthalazine-5,10-dione carbonitriles and carboxylates. Journal of Molecular Catalysis A: Chemical. 2012;**361-362**:58-67. DOI: 10.1016/j.molcata.2012.05.003

[15] Zhan W, Guo Y, Wang Y, Guo Y, Lu G. Synthesis of lathanum or La-B doped KIT-6 mesoporous materials and their application in the catalytic oxidation of styrene. Journal of Rare Earths. 2010;**28**:369-375. DOI: 10.1016/S1002-0721(09)60105-8

[16] Filip M, Todorova S, Shopska M, Ciobanu M, Papa F, Somacescu S, et al. Effects of Ti loading on activity and redox behavior of metals in PtCeTi/KIT-6 catalysts for CH_4 and CO oxidation. Catalysis Today. 2018;**306**:138-144. DOI: 10.1016/j.cattod.2017.02.013

[17] Ghohe NM, Tayebee R, Amini MM. Synthesis and characterization of mesoporous NbZr/KIT-6 as a productive catalyst for the synthesis of benzylpyrazolyl coumarins. Materials Chemistry and Physics. 2019;**223**:268-276. DOI: 10.1016/j.matchemphys.2018.10.067

[18] He F, Luo J, Liu S. Novel metal loaded KIT-6 catalysts and their applications in the catalytic combustion of chlorobenzene. Chemical Engineering Journal. 2016;**294**:362-370. DOI: 10.1016/j.cej.2016.02.068

[19] Liu C-C, Lin T-S, Chan SI, Mou C-Y. A room temperature catalyst for toluene aliphatic C–H bond oxidation: Tripodal tridentate copper complex immobilized in mesoporous silica. Journal of Catalysis. 2015;**322**:139-151. DOI: 10.1016/j.jcat.2014.12.005

[20] Mureşeanu M, Pârvulescu V, Ene R, Cioateră N, Pasatoiu TD, Andruh M. Cu(II) complexes imobilized on functionalized mesoporous silica as catalysts for biomimetic oxidations. Journal of Materials Science. 2009;**44**:6795-6804. DOI: 10.1007/s10853-009-3682-6

[21] Parvulescu V, Anastasescu C, Constantin C, Su B-L. Mono (V, Nb) or bimetallic (V–Ti, Nb–Ti) ions modified MCM-41 catalysts: Synthesis, characterization and catalysis in oxidation of hydrocarbons (aromatics and alcohols). Catalysis Today. 2003;**78**:477-485. DOI: 10.1016/S0920-5861(02)00330-9

[22] Amama PB, Lim S, Ciuparu D, Pfefferle L, Haller GL. Hydrothermal synthesis of MCM-41 using different ratios of colloidal and soluble silica. Microporous and Mesoporous Materials. 2005;**81**:191-200. DOI: 10.1016/j.micromeso.2005.02.001

[23] Choi KM, Tatsumi T, Yokoi T, Kuroda K. Usefulness of alkoxyltitanosiloxane for the preparation of mesoporous silica containing a large amount of isolated titanium. Journal of Colloid and Interface Science. 2011;**359**:240-247. DOI: 10.1016/j.jcis.2011.03.016

[24] Parvulescu V, Tablet C, Anastasescu C, Su B-L. Activity and stability of bimetallic Co (V, Nb, La)-modified MCM-41 catalysts. Catalysis Today. 2004;**93-95**:307-313. DOI: 10.1016/j.cattod.2004.06.006

[25] Todorova S, Pârvulescu V, Kadinov G, Tenchev K, Somacescu S, Su B-L. Metal states in cobalt- and cobalt-vanadium-modified MCM-41 mesoporous silica catalysts and their

activity in selective hydrocarbons oxidation. Microporous and Mesoporous Materials. 2008;**113**:22-30. DOI: 10.1016/j.micromeso.2007.10.047

[26] Szegedi Á, Kónya Z, Méhnb D, Solymár E, Pál-Borbély G, Horváth ZE, et al. Spherical mesoporous MCM-41 materials containing transition metals: Synthesis and characterization. Applied Catalalysis A: General. 2004;**272**:257-266. DOI: 10.1016/j.apcata.2004.05.057

[27] Yang YH, Lim S, Du GA, Chen Y, Ciuparu D, Haller GL. Synthesis and characterization of highly ordered Ni-MCM-41 mesoporous molecular sieves. The Journal of Physical Chemistry B. 2005;**109**:13237-13246. DOI: 10.1021/jp044227i

[28] Petcu G, Anghel EM, Somacescu S, Preda S, Culita DC, Mocanu S, et al. Hierarchical zeolite Y containing Ti and Fe oxides as photocatalysts for degradation of amoxicillin. Journal of Nanoscience and Nanotechnology. 2020;**20**:1158-1169. DOI: 10.1166/jnn.2019.16981

[29] Kolev H, Todorova S, Naydenov A, Ene R, Ivanov G, Parvulescu V, et al. Catalytic activity of mesoporous SBA-15 modified with Pt and Ti in a deep methane, n-hexane and CO oxidation. Athens Journal of Sciences. 2014;**1**:9-20. DOI: 10.30958/ajs.1-1-1

[30] Cánepa AL, Elías VR, Vaschetti VM, Sabre EV, Eimer GA, Casuscelli SG. Selective oxidation of benzyl alcohol through eco-friendly processes using mesoporous V-MCM-41, Fe-MCM-41 and Co-MCM-41 materials. Applied Catalysis A: General. 2017;**545**:72-78. DOI: 10.1016/j.apcata.2017.07.039

[31] Baoshan KW, Han LC, Liu J. Synthesis, characterization of MCM-41 with high vanadium content in the framework and its catalytic performance on selective oxidation of cyclohexane. Applied Catalysis A: General. 2014;**479**:70-75. DOI: 10.1016/j.apcata.2014.04.004

[32] Shylesh S, Singh AP, Shylesh S, Singh AP. Synthesis, characterization, and catalytic activity of vanadium-incorporated, -grafted, and -immobilized mesoporous MCM-41 in the oxidation of aromatics. Journal of Catalysis. 2004;**228**:333-346. DOI: 10.1016/j.jcat.2004.08.037

[33] Kondratenko EV, Cherian M, Baerns M, Su D, Schlögl R, Wang X, et al. Oxidative dehydrogenation of propane over V/MCM-41 catalysts: Comparison of O_2 and N_2O as oxidants. Journal of Catalysis. 2005;**234**:131-142. DOI: 10.1016/j.jcat.2005.05.025

[34] Jha RK, Shylesh S, Bhoware SS, Singh AP. Oxidation of ethyl benzene and diphenyl methane over ordered mesoporous M-MCM-41 (M = Ti, V, Cr): Synthesis, characterization and structure–activity correlations. Microporous and Mesoporous Materials. 2003;**95**:154-163. DOI: 10.1016/j.micromeso.2006.04.018

[35] Ziolek M, Nowak I. Characterization techniques employed in the study of niobium and tantalum-containing materials. Catalysis Today. 2003;**78**:543-553. DOI: 10.1016/S0920-5861(02)00353-X

[36] Golinska H, Decyk P, Ziolek M, Kujawa J, Filipek E. Sb-V-Ox catalysts—Role of chemical composition of MCM-41 supports in physicochemical properties. Catalysis Today. 2009;**142**:175-180. DOI: 10.1016/j.cattod.2008.10.044

[37] Braganca LFFPG, Ojedac M, Fierroc JLG, Pais da Silva MI. Bimetallic Co–Fe nanocrystals deposited on SBA-15 and HMS mesoporous silicas as catalysts for Fischer–Tropsch synthesis. Applied Catalysis A: General. 2012;**423-424**:146-153. DOI: 10.1016/j.apcata.2012.02.031

[38] Nilsen MH, Antonakou E, Bouzga A, Lappas A, Mathisen K, Stocker M. Investigation of the effect of metal sites in Me–Al-MCM-41 (Me = Fe, Cu or Zn) on the catalytic behavior during the pyrolysis of wooden based biomass. Microporous and Mesoporous Materials. 2007;**105**:189-203. DOI: 10.1016/j.micromeso.2007.05.059

[39] Parida KM, Dash SK. Adsorption of Cu^{2+} on spherical Fe-MCM-41 and its application for oxidation of adamantane. Journal of Hazardous Materials. 2010;**179**:642-649. DOI: 10.1016/j.jhazmat.2010.03.051

[40] Kucherov AV, Shigapov AN, I vanov AV, Kucherova TN, Kustov LM. Distribution and properties of catalytically active Cu^{2+}-sites on a mesoporous MCM-41 silicate modified by Al, Zr, W, B, or P ions. Catalysis Today. 2005;**110**:330-338. DOI: 10.1016/j.cattod.2005.09.031

[41] Laha SC, Mukherjee P, Sainkar SR, Kumar R. Cerium containing MCM-41-type mesoporous materials and their acidic and redox catalytic properties. Journal of Catalysis. 2002;**207**:213-223. DOI: 10.1006/jcat.2002.3516

[42] Dai Q, Wang X, Chen G, Zheng Y, Lu G. Direct synthesis of cerium (III)-incorporated SBA-15 mesoporous molecular sieves by two-step synthesis method. Microporous and Mesoporous Materials. 2007;**100**:268-275. DOI: 10.1016/j.micromeso.2006.11.015

[43] Derylo-Marczewska A, Gacb W, Popivnyak N, Zukocinski G, Pasieczna S. The influence of preparation method on the structure and redox properties of mesoporous Mn-MCM-41 materials. Catalysis Today. 2006;**114**:293-306. DOI: 10.1016/j.cattod.2006.02.066

[44] Wu H-Y, Zhang X-L, Yang C-Y, X. Chen X, Zheng X-C. Alkali-hydrothermal synthesis and characterization of W-MCM-41 mesoporous materials with various Si/W molar ratios. Applied Surface Science. 2013;270:590-595. Doi: 10.1016/j.apsusc.2013.01.090

[45] Parvulescu V, Ciobanu M, Petcu G, Anghel EM, Tungsten–modified SB-L. HMS catalysts for high selective oxidation of styrene with aqueous hydrogen peroxide. Revue Roumaine de Chimie. 2018;**63**:847-853

[46] Shah P, Ramaswamy AV, Lazar K, Ramaswamy V. Synthesis and characterization of tin oxide-modified mesoporous SBA-15 molecular sieves and catalytic activity in trans-esterification reaction. Applied Catalysis A: General. 2004;**273**:239-302. DOI: 10.1016/j.apcata.2004.06.039

[47] Kawi S, Liu SY, Shen S-C. Catalytic decomposition and reduction of N_2O on Ru/MCM-41 catalyst. Cataysis Today. 2001;**68**:237-244. DOI: 10.1016/S0920-5861(01)00283-8

[48] Jang M, Park JK, Shin EW. Lanthanum functionalized highly ordered mesoporous media: Implications of arsenate removal. Microporous and Mesoporous Materials. 2004;**75**:159. DOI: 10.1016/j.micromeso.2004.05.018

[49] Parvulescu V, Anastasescu C, Su BL. Bimetallic Ru-(Cr, Ni, or Cu) and La-(Co or Mn) incorporated MCM-41 molecular sieves as catalysts for oxidation of aromatic hydrocarbons. Journal of Molecular Catalysis A: Chemical. 2004;**211**:143-148. DOI: 10.1016/j.molcata.2003.10.011

[50] Parvulescu V, Constantin C, Su B-L. Liquid phase oxidation of aromatic hydrocarbons using highly ordered Nb and NbCo-MCM-41 nanoreactors. Journal of Molecular Catalysis A: Chemical. 2003;**202**:171-178. DOI: 10.1016/S1381-1169(03)00202-4

[51] Zhou S, Yang F, Wang B, Su H, Lu K, Din Y, et al. Oriented decoration in metal-functionalized ordered mesoporous silicas and their catalytic applications in the oxidation of aromatic compounds. Catalysts. 2018;**8**:80-110. DOI: 10.3390/catal8020080

[52] Prabhu A, Kumaresan L, Palanichamy M, Murugesan V. Cerium-incorporated cage-type mesoporous KIT-6 materials: Synthesis, characterization and catalytic applications. Applied Catalysis A: General. 2010;**374**:11-17. DOI: 10.1016/j.apcata.2009.11.016

[53] Xiong G, Cao Y, Guo Z, Jia Q, Tian F, Liu L. The roles of different titanium species in TS-1 zeolite in propylene epoxidation studied by in situ UV Raman spectroscopy. Physical Chemistry Chemical Physics. 2016;**18**:190-196. DOI: 10.1039/c5cp05268h

[54] Zhao W, Li G, Tang Z. Metal-organic frameworks as emerging platform for supporting isolated single-site catalysts. Nano Today. 2019;**27**:178-197. DOI: 10.1016/j.nantod.2019.05.007

[55] Roy SK, Dutta D, Talukdar AK. Highly effective methylated Ti MCM-41 catalyst for cyclohexene oxidation. Materials Research Bulletin. 2018;**103**:38-46. DOI: 10.1016/j.materresbull.2018.03.017

[56] Orbeci C, Stănescu R, Negoescu D, Parvulescu V. Synthesis, characterization and functionalization of MCM-41 for the removal of organic compounds from waste waters. Environmental Engineering and Management Journal. 2017;**16**:*553-560*

[57] Boulaoued A, Fechete I, Donnio B, Bernard M, Turek F, Garin F. Mo/KIT-6, Fe/KIT-6 and Mo–Fe/KIT-6 as new types of heterogeneous catalysts for the conversion of MCP. Microporous and Mesoporous Materials. 2012;**155**:131-142. DOI: 10.1016/j.micromeso.2012.01.028

[58] Venezia AM, Carlo GG, Liotta LF, Pantaleo G, Kantcheva M. Effect of Ti(IV) loading on CH_4 oxidation activity and SO_2 tolerance of Pd catalysts supported on silica SBA-15 and HMS. Applied Catalysis B: Environmental. 2011; 106:529-539. Doi: 10.1016/j.apcatb.2011.06.013

Chapter 3

Copolymer Synthesis with Redox Polymerization and Free Radical Polymerization Systems

Melahat Göktaş

Abstract

In this study, block copolymer synthesis was evaluated by combining the redox polymerization technique and different polymerization techniques. By combining such different polymerization techniques, block copolymer synthesis has recently become an important part of polymer synthesis and polymer technology. The block/graft copolymers synthesized by combining such different techniques contribute greatly to macromolecular engineering. In today's polymer synthesis, copolymer synthesis is of great interest in polymer technologies especially by using controlled radical polymerization and different polymerization techniques together. The success achieved in copolymer synthesis by using different polymerization techniques such as ATRP-ROP, RAFT-ROP, and ATRP-RAFT on the same step or different steps was achieved by combining redox polymerization with moderate polymerization conditions and controlled radical polymerization techniques as well.

Keywords: redox systems, free controlled radical polymerization, redox polymerization, copolymer, macroinitiator

1. Introduction

Since the 1980s, about 50% of the total production of synthetic polymers used as plastics worldwide has been achieved through free radical polymerization. Peroxy compounds in technical polymerization processes have played the most important role in addition to 60-year redox systems and azo initiators for nearly 100 years. For nearly 30 years, polymer synthesis with free radical polymerization reactions has attracted considerable attention technically, even though their share in total polymer production is still quite small [1].

The most important advantage of conventional free radical polymerization, which is widely used, is that many monomers can be polymerized using this method and that this polymerization can be made under moderate conditions. The most important disadvantage of this polymerization technique is that the polymer architecture and molecular weight are not controlled and also the production of polymers with large molecular weight distribution [2]. The manufacturing of polymers, of which molecular architecture and molecular weight can be controlled in recent years and which have low molecular weight distribution (polydispersity), has been made possible with controlled radical polymerization techniques. Under favor of controlled radical polymerization techniques, polymers with narrow molecular weight distribution can be produced in a desired molecular weight and desired molecular way in a controlled and repeatable manner. Synthesis of polymers, which have the star, comb, brush, worm, or graft architecture, is provided by molecular structure and size-controlled radical polymerization techniques [3–7].

Until today, the synthesis of block copolymers has usually been made through ionic polymerization. But ionic polymerization requires strict conditions, and the number of monomers is relatively limited. To overcome these disadvantages, simpler and easier techniques have been used recently for block copolymer synthesis [8, 9]. It has been possible to be successful in block copolymer synthesis in recent years with RAFT-ROP [10], ATRP-ROP [11], and redox polymerization-ATRP methods which have many advantages compared to other popular methods and have been implemented by using different techniques together [12]. Due to the practicability of two transformations at the same time or through separate steps, it minimizes homopolymerization which causes side reactions. Combining different polymerization techniques should be an interesting method for block and graft copolymers because the presence of more than one monomer in a polymer chain has been by combining such different techniques [10, 13–15]. The new polymers may have amazing features with their various compositions and architectures. The synthesis of block and graft copolymers was successfully performed by combining controlled radical polymerization techniques and redox polymerization [16]. The synthesis of block copolymers ends with traditional radical polymerization based on the connection of functional groups of the chain and polymers. Though this strategy was effective and successful, it was difficult to test the molecular weight and architecture of the polymer which was obtained. To be able to solve this problem, controlled free radical polymerization techniques were developed quickly [17].

In this present study, the synthesis of block copolymers over separate steps or on the same step was examined with different free radical polymerization techniques and redox polymerization methods. Copolymer synthesis by combining such different techniques has recently attracted considerable attention in polymer synthesis science.

2. Redox polymerization

Redox initiator polymerization was first discovered in Germany (1937); then it attempted to remove the induction period in aqueous or emulsion polymerization by adding a reducing agent to the oxidant initiator in the USA (1945) and in England (1946). Only the increase of rate of polymerization (RP) along with the expected decrease in the induction period was observed at that rate. The main characteristic of the compounds which form a redox pair for aqueous polymerization is their solubility in water, producing active, stable, and relatively fast radicals [1, 18].

The polymerizations activated by a reaction between an oxidant and a reducing agent are called redox polymerizations. The essence of redox activation is a reduction-oxidation process. In this process, an oxidant, i.e., Ce (IV) or Mn (III), forms a complex by simply reacting organic molecules at the beginning, which then decomposes unimolecularly to produce free radicals that initiate polymerization. There are peroxides, persulfates, peroxide phosphate, and salts of transition metals among the oxidants commonly used. These oxidants form effective redox systems with various reducing agents such as alcohols, aldehydes, amines, and thiols for the aqueous polymerization of vinyl monomers. The basic properties of the components forming a redox pair for aqueous polymerization are their water solubility and the quite rapid and stable release of active radicals [19, 20]. It is easy to control the reaction rate by changing the concentration of metal ion or peroxide, except for the use of low temperatures in redox systems [21]. There are many studies about block copolymer synthesis in the literature. Starting with a redox operation is only one method to obtain such polymers [22, 23].

The synthesis of block copolymers with redox systems provides a number of technical and theoretical advantages as compared with the other methods. Redox polymerization minimizes side reactions under favor of its applicability at low temperatures [24]. In radical polymerization, redox systems are widely used as initiators, and a result is accomplished in a very short time. When compared with the other methods, it is the main advantage of processing at a very moderate temperature (low; 30 kcal/mole for thermal start and 10–20 kcal/mole for activation energy). This shows that it can minimize possible side reactions. The Ce(IV) or permanganate initiators, which combine with a reducing agent that includes a hydroxyl or carboxyl group, are more commonly used initiators [25].

The mechanism and the speed of redox polymerization can be shown with the following equations:

For the first radical formation,

$$Ce(IV) + R \xrightarrow{k_d} R\cdot + Ce(III) + H^+$$

where R·is the form of one or two CH_2OH functional groups converted to CH_2O and k_d is the rate constant for initiator cleavage in the redox reaction.

For initiation,

$$R\cdot + M \xrightarrow{k_i} RM_1\cdot$$

where M in the equation shows the polymerizable monomer by the redox method and k_i shows the starting rate constant.

For growing,

$$\begin{array}{c} RM_1^{\cdot} + M \xrightarrow{k_p} RM_2^{\cdot} \\ \vdots \qquad\qquad \vdots \\ RM_{n-1}^{\cdot} + M \xrightarrow{k_p} RM_n^{\cdot} \end{array}$$

There could be three types of endings: linear, bimolecular, and oxidative termination of the first radical:

For linear termination,

$$RM_n^{\cdot} + Ce(IV) \xrightarrow{k_{t1}} RM_n + Ce(III) + H^+$$

where kt_1 is the linear termination rate constant.

For bimolecular termination,

$$RM_n^{\cdot} + RM_m^{\cdot} \xrightarrow{k_{t2}} RM_n + mR\cdot$$

where kt_2 is the bimolecular termination rate constant.

For oxidative termination of the first radical,

$$R + Ce(IV) \xrightarrow{k_o} \text{organic products} + Ce(III) + H^+$$

where k_o is the rate constant of the termination of the first radical.

3. Catalysts used in redox polymerization systems

Various compounds such as ceric, manganese, copper, iron, vanadium ion salts, and hydrogen peroxide were used as catalysts for the synthesis of block copolymers through redox polymerization. The ceric-based catalysts are the most widely used in these catalyst systems. The redox systems containing other catalysts were also examined. Ce(IV) or permanganate initiators, which combine with reducing agent containing a hydroxyl or carboxyl group, are the more commonly used initiators.

Ceric salts have shapes such as ceric(IV) ammonium nitrate (CAN), ceric(IV) ammonium sulfate (CAS), ceric(IV) sulfate (CS), and ceric perchlorate. As oxidation strength, ceric perchlorate > ceric nitrate > ceric sulfate were observed (1.7, 1.6, and 1.4 V), respectively, in the studies carried out with vinyl monomers [26]. A wide range of usages in the free radical production has been found by taking advantage of its amplifying properties in redox polymerization. The reduction reaction is given below.

$$Ce(IV) + e^- \longrightarrow Ce(III)$$

The Ce(IV) salts and the Ce(IV) salt-reducing substance system are used as initiators for vinyl polymerizations in aqueous acidic solutions [27]. Organic reductant substances most commonly used with the Ce(IV) salts are alcohols, glycols, aldehydes, ketones, and carboxylic acids [27, 28]. Ce(IV) salts are used only in acidic solutions and most preferably in 0.5 or higher acid concentrations [29]. The solution's color is yellow. The turning point can be determined even without an indicator in hot and non-dilute solutions.

It has been proven by research that Ce(IV) ion cannot initiate acrylamide polymerization alone and water is not oxidized by Ce(IV) ions [27]. So the radicals that start polymerization occur as a result of the reaction between the Ce(IV) ion and the reducing substance. A general mechanism is proposed for this.

$$R\,Ce(IV) + e^- \longrightarrow \text{polymer} + Ce(III) + H^+$$

When keeping the concentration of methyl methacrylate and Azo I constant and increasing the concentration of Ce(IV) up to 6×10^{-4} mol L^{-1}, the polymerization rate also increased proportionally with $[Ce(IV)]^{1/2}$ in the methyl methacrylate polymerization initiated by the hydroxyl functional group with a redox pair Ce(IV)-Azo I. This adherence explains the bimolecular termination. The rapid degradation of polymerization in high Ce(IV) concentrations indicates that active chains are terminated by Ce(IV) [30].

Arslan and Hazer [31] reported the polymerization of methyl methacrylate initiated by ceric ammonium nitrate (MMA) in the form of combination with polytetrafuran diol (PTHF-diol) and polycaprolactone diol (PCL-diol) in aqueous nitric acid. PMMA-*b*-PTHF and PMMA-*b*-PCL block copolymers were obtained. The polymerization reactions are presented in **Figure 1**.

Hazer et al. [32] searched the polymerization of methyl methacrylate initiated by ceric ammonium nitrate and poly(glycidyl azide)-diol in the aqueous nitric acid. Poly(methyl methacrylate)-*b*-poly(glycidyl acrylate) copolymer was obtained. The reaction mechanism is shown in **Figure 2**.

Çakmak et al. [33] used redox reactions in the preparation of acrylamide-ethylene glycol block copolymers (PAAm-PEG) containing azo groups in the main chain. The synthesis pathway of the copolymers is shown in **Figure 3**.

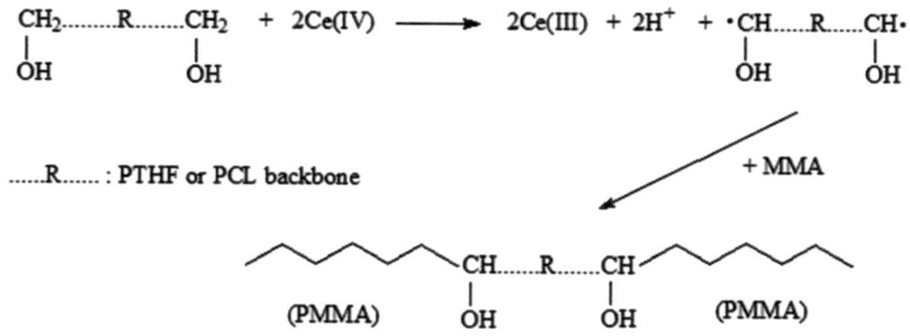

Figure 1.
Synthesis of PMMA-b-PCL-b-PMMA block copolymer with PCL-diol/Ce(IV) redox systems.

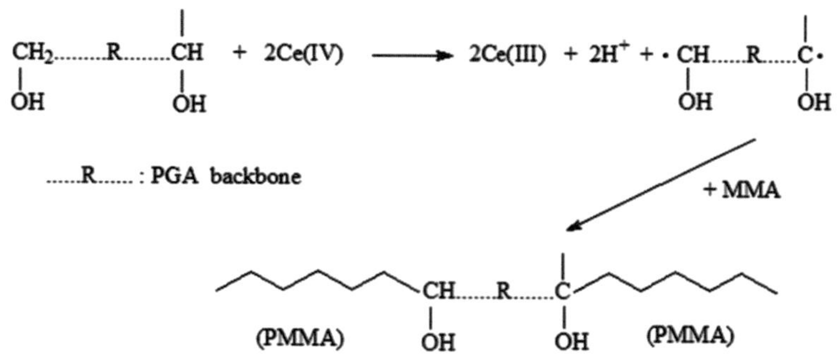

Figure 2.
Polymerization of methyl methacrylate initiated by ceric ammonium nitrate in combination with poly(glycidyl azide)-diol (PGA-diol).

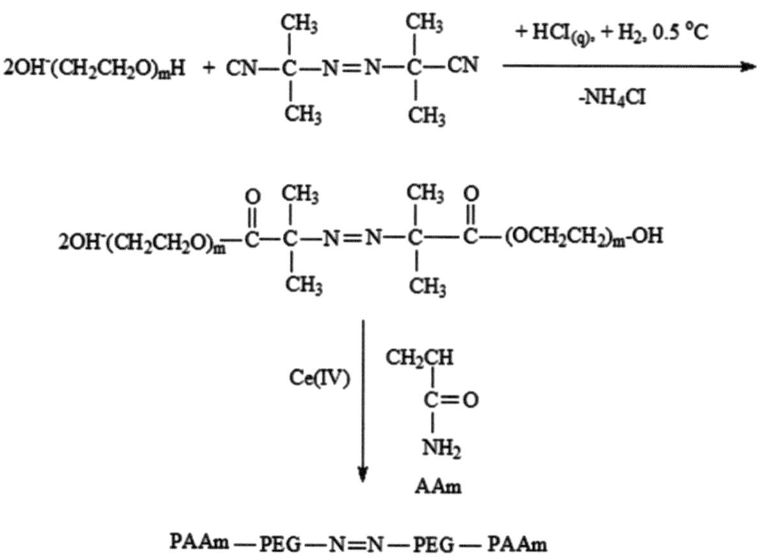

Figure 3.
Polymerization of acrylamide with poly(ethylene glycol)azoester/Ce^{+4} redox system.

Shimizu et al. [34] synthesized redox reaction with the poly(*N*-isopropylacrylamide-*b*-ethylene glycol) [(PNIPAM)-*b*-(PEG)] thermo-responsive block copolymers in ceric ammonium nitrate catalyzer using the PEG macroinitiator. The synthesis mechanism of block copolymers is shown in **Figure 4**.

Göktaş et al. [12] evaluated poly(methyl methacrylate)-*b*-poly(*N*-isopropylacrylamide) [PMMA-*b*-PNIPAM] block copolymers in two steps under the catalyzer of ceric ammonium (IV) nitrate (CAN) [$Ce(NH_4)_2(NO_3)_6$] by using 3-bromo-1-propanol initiator, suitable for both redox polymerization and atom transfer radical polymerization which is one of the controlled radical polymerization techniques. The synthesis mechanisms of the polymerization are shown in **Figures 5** and **6**.

Zhuang et al. [16] evaluated poly(hydroxylethyl methacrylate)-branched-poly (acrylamide) (PHEMA-*branched*-PAM) polymer by combining atom transfer radical and redox polymerization methods. The synthesis mechanism of the polymer is shown in **Figure 7**.

Göktaş et al. [35] evaluated poly(methyl methacrylate-b-styrene) and poly(methyl methacrylate-b-acrylamide) which were synthesized in two steps using a combination of the redox polymerization method and the atom transfer radical polymerization (ATRP) method. The synthesis mechanisms of the polymerization are shown in **Figures 8** and **9**.

Çakmak et al. [24] evaluated poly(acrylonitrile)-*block*-poly(ethylene glycol) block copolymer via redox polymerization using Mn(III) as catalyzer. The synthesis pathway of the copolymers is shown in **Figure 10**.

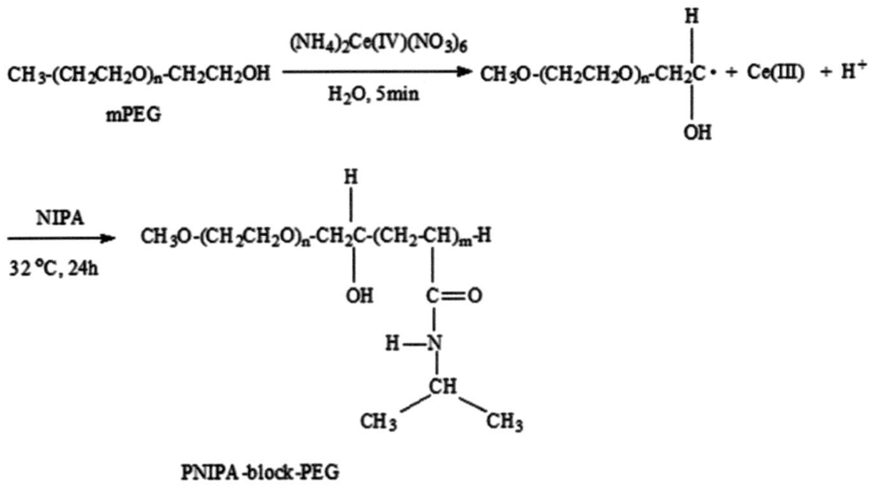

Figure 4.
*Synthesis of poly(*N*-isopropylacrylamide)-*block*-poly(ethylene glycol) block copolymer via poly(ethylene glycol)/Ce(IV) redox pair.*

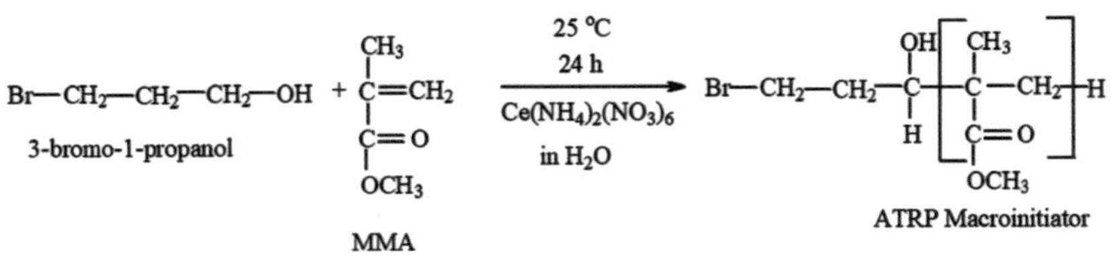

Figure 5.
Reaction pathways in the synthesis of ATRP macroinitiator.

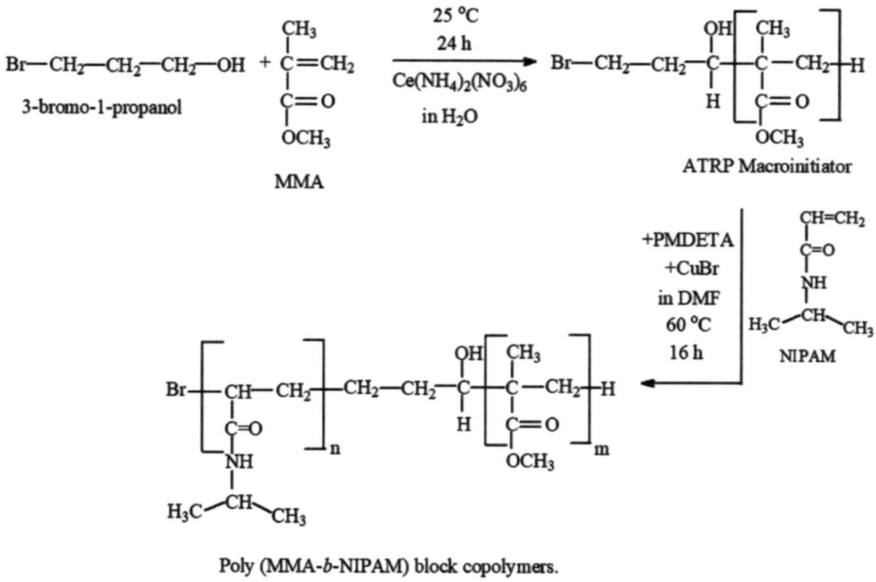

Figure 6.
*Reaction pathways in the synthesis of PMMA-*b*-PNIPAM block copolymers.*

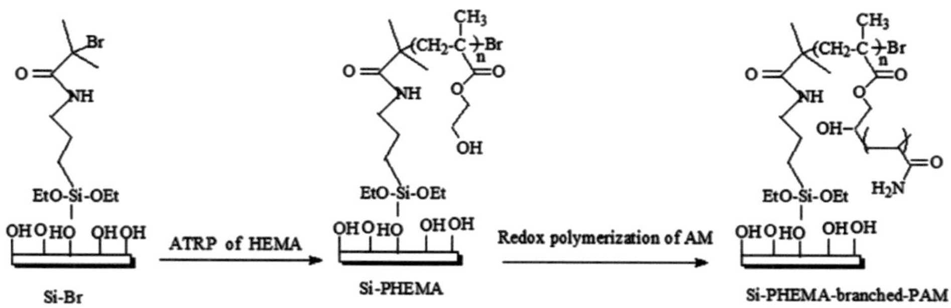

Figure 7.
*The synthesis route of PHEMA-*branched*-PAM layers via ATRP and redox polymerization on silicon substrates.*

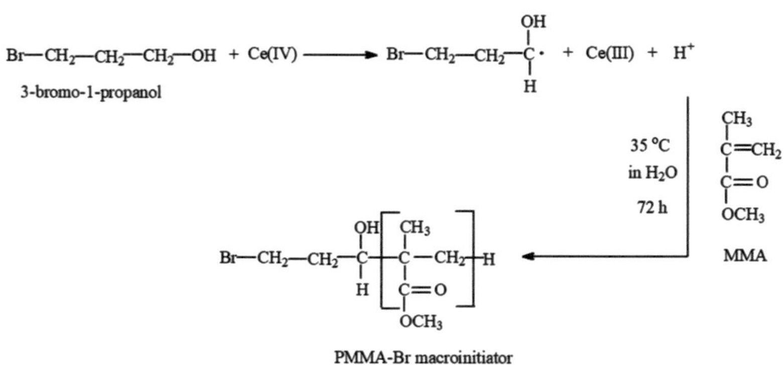

Figure 8.
***Figure 1** Chemical synthesis of PMMA-Br macroinitiator via redox polymerization.*

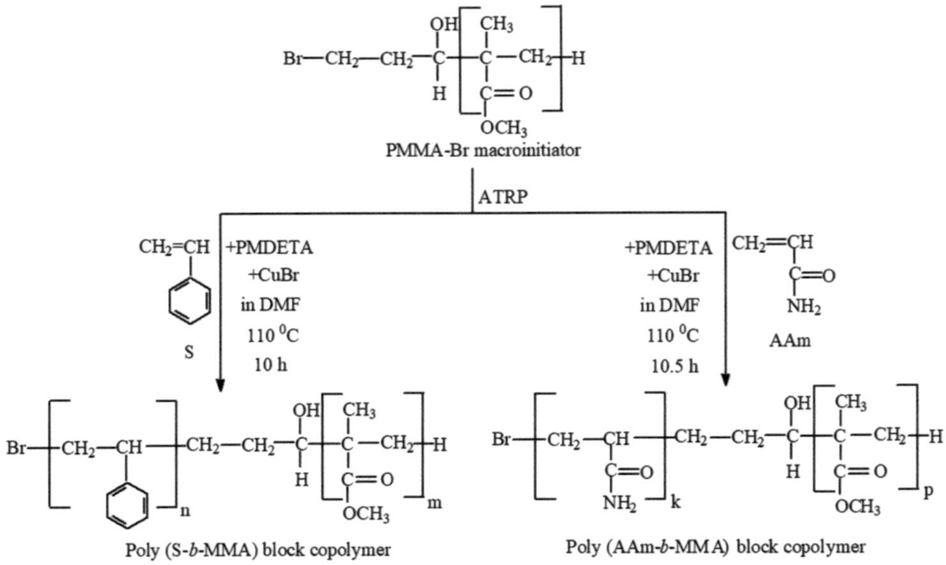

Figure 9.
*Synthetic route poly(MMA-*b*-S) and poly(MMA-*b*-AAm) for block copolymers.*

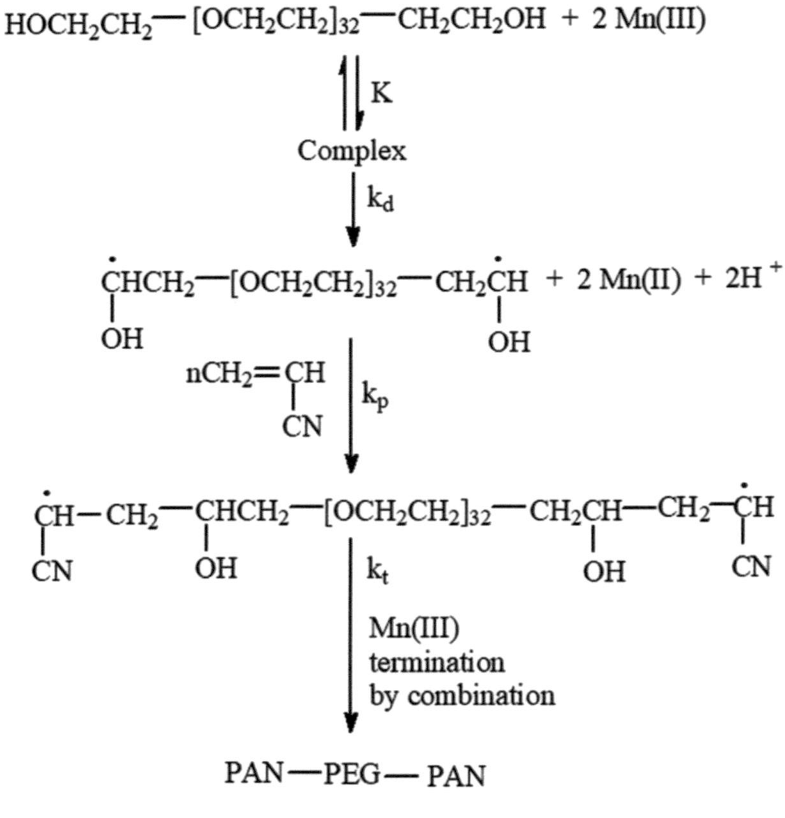

Figure 10.
*Synthesis of poly(acrylonitrile)-*block*-poly(ethylene glycol) block copolymer via poly(ethylene glycol)/Mn(III) redox couple.*

Liu et al. [36] evaluated methyl acrylate (MA) and poly(ethylene glycol) (PEG) block copolymers using a novel redox system-potassium diperiodatocuprate(III)

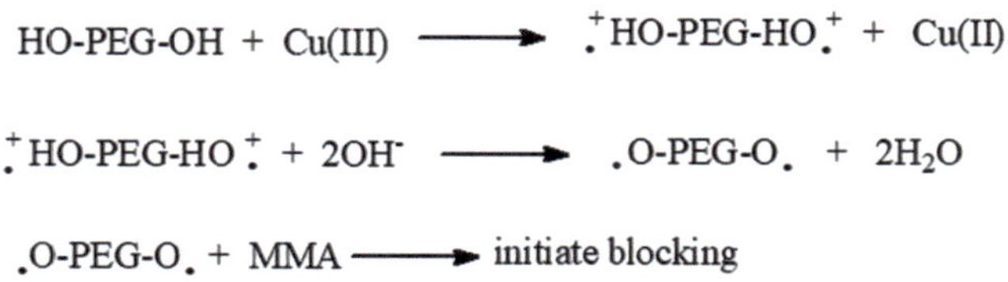

Figure 11.
Block copolymerization of methyl acrylate (MA) and poly(ethylene glycol) (PEG) using potassium diperiodatocuprate(III)[DPC]/PEG redox system.

[DPC]/PEG system in alkaline aqueous medium. The synthesis mechanism of the polymer is shown in **Figure 11**.

4. Conclusion

Today, polymer materials science dominates the synthesis and design of polymers with complex architecture and advanced properties. The functional copolymers with block, graft, star, and brush structures can be prepared by controlled radical polymerization techniques. Copolymer synthesis has been important recently, especially by using controlled radical polymerizations in combination with traditional polymerization methods such as cationic polymerization and redox polymerization. This is because the homopolymer formation is minimized in block copolymer synthesis with the combination of such different techniques.

In this study, it was emphasized that block copolymer synthesis has superior properties compared to traditional polymerization methods using the redox polymerization method and different polymerization techniques, because combining different monomers in the same polymer chain in copolymer synthesis with multi-synthesis methods contributes positively to polymer material science.

Author details

Melahat Göktaş
Department of Science Education, Yüzüncü Yil University, Van, Turkey

*Address all correspondence to: melahat_36@hotmail.com

IntechOpen

© 2019 The Author(s). Licensee IntechOpen. This chapter is distributed under the terms of the Creative Commons Attribution License (http://creativecommons.org/licenses/by/3.0), which permits unrestricted use, distribution, and reproduction in any medium, provided the original work is properly cited.

References

[1] Braun D. Origins and development of initiation of free radical polymerization processes. International Journal of Polymer Science. 2009;**2009**:10. DOI: 10.1155/2009/893234

[2] Misra GS, Bhattacharya SN. Polymerization of acrylamide initiated by the Ce^{4+}/thiourea redox system. Journal of Polymer Science: Polymer Chemistry Edition Banner. 1982;**8**:61-131. DOI: 10.1002/pol.1982.170200114

[3] Okada S, Matyjaszewski K. Synthesis of bio-based poly(N-phenylitaconimide) by atom transfer radical polymerization. Journal of Polymer Science: Polymer Chemistry Edition Banner. 2015;**53**:822-827. DOI: 10.1002/pola.27507

[4] Lee IH, Discekici EH, Shankel SL, Anastasaki A, De Alaniz JR, Hawker CJ, et al. Desulfurization–bromination: Direct chain-end modification of RAFT polymers. Polymer Chemistry. 2017;**8**:7188-7194. DOI: 10.1039/c7py01702b

[5] Zaremski M, Eremeev I, Garina E, Borisova O, Korolev B. Controlled synthesis of random, block-random and gradient styrene methyl methacrylate and acrylonitrile terpolymers via nitroxide-mediated free radical polymerization. Journal of Polymer Research. 2017;**24**:151. DOI: 10.1007/s10965-017-1303-7

[6] Diaz-Silvestre S, Saldivar-Guerra E, Rivera-Vallejo C, St Thomas C, Cabello-Romero J, Guerrero-Santos R, et al. Synthesis of associative block copolymers electrolytes via RAFT polymerization. Polymer Bulletin. 2018;**75**:891-907. DOI: 10.1007/s00289-017-2071-6

[7] Chiefari J, Chong YK, Ercole F, Krstina J, Jeffery J, Le TPT, et al. Living free-radical polymerization by reversible addition-fragmentation chain transfer: The RAFT process. Macromolecules. 1998;**31**:5559-5562. DOI: 10.1021/ma9804951

[8] Öztürk T, Göktaş M, Savaş B, Işıklar M, Atalar MN, Hazer B. Synthesis and characterization of poly(vinylchloride-*graft*-2-vinylpyridine) graft copolymers using a novel macroinitiator by reversible addition-fragmentation chain transfer polymerization. E-Polymers. 2014;**14**:27-34. DOI: 10.1515/epoly-2013-0011

[9] Zhao M, Shi Y, Fu Z, Yang W. Preparation of PMMA-b-PSt block copolymer via seeded emulsion polymerization in the presence of 1,1-diphenylethylene. Macromolecular Reaction Engineering. 2014;**8**:555-563. DOI: 10.1002/mren.201300174

[10] Göktaş M, Öztürk T, Atalar MN, Tekeş AT, Hazer B. One-step synthesis of triblock copolymers via simultaneous reversible-addition fragmentation chain transfer (RAFT) and ring-opening polymerization using a novel difunctional macro-RAFT agent based on polyethylene glycol. Journal of Macromolecular Science, Part A. 2014;**51**:854-863. DOI: 10.1080/10601325.2014.953366

[11] Öztürk T, Yavuz M, Göktaş M, Hazer B. One-step synthesis of triarm block copolymers by simultaneous atom transfer radical and ring-opening polymerization. Polymer Bulletin. 2016;**73**:1497-1513. DOI: 10.1007/s00289-015-1558-2

[12] Göktaş M, Deng G. Synthesis of poly(methyl methacrylate)-b-poly(N-isopropylacrylamide) block copolymer by redox polymerization and atom transfer radical polymerization. Indonesian Journal of Chemistry. 2018;**18**:537-543. DOI: 10.22146/ijc.28645

[13] Öztürk T, Göktaş M, Hazer B. One-step synthesis of triarm block copolymers via simultaneous reversible-addition fragmentation chain transfer and ring-opening polymerization. Journal of Applied Polymer Science. 2010;**117**:1638-1645. DOI: 10.1002/app.32031

[14] Öztürk T, Göktaş M, Hazer B. Synthesis and characterization of poly(methyl methacrylate-block-ethylene glycol-block-methyl methacrylate) block copolymers by reversible addition-fragmentation chain transfer polymerization. Journal of Macromolecular Science, Part A—Pure and Applied Chemistry. 2011;**48**:65-72. DOI: 10.1080/10601325.2011.528310

[15] Öztürk T, Kayğın O, Göktaş M, Hazer B. Synthesis and characterization of graft copolymers based on polyepichlorohydrin via reversible addition-fragmentation chain transfer polymerization. Journal of Macromolecular Science, Part A—Pure and Applied Chemistry. 2016;**53**:362-367. DOI: 10.1080/10601325.2016.1166002

[16] Zhuang D, Shen H, Liu G, Yu C, Yang JA. Combining signal amplification of atom transfer radical polymerization and redox polymerization for visual biomolecules detection. Journal of Polymer Science Part A: Polymer Chemistry. 2014;**52**:2791-2799. DOI: 10.1002/pola.27303

[17] Jones GR, Li Z, Anastasaki A, Lloyd DJ, Wilson P, Zhang Q, et al. Rapid synthesis of well-defined polyacrylamide by aqueous Cu(0)-mediated reversible-deactivation radical polymerization. Macromolecules. 2016;**49**:483-489. DOI: 10.1021/acs.macromol.5b01994

[18] Herman FM, Norman GG, Norbert MB. Encyclopedia of Polymer Science and Technology. New York: John Wiley & Sons; 1969. p. 10

[19] Cakmak I. Synthesis of block copolymers by redox macro initiators. Journal of Macromolecular Science, Part A. 1995;**A32**:197-206. DOI: 10.1080/10601329508020328

[20] Öztürk T, Çakmak İ. Synthesis of block copolymers via redox polymerization process: A critical review. Iranian Polymer Journal. 2007;**16**:561-581

[21] Stewens PM. Polymer Chemistry: An Introduction. 3rd ed. New York: Oxford University; 1999. p. 550

[22] Nagarajan S, Srinivason KSV. Redox polymerization process: An efficient tool for the synthesis of block copolymers. European Polymer Journal. 1994;**30**:113-119. DOI: 10.1002/pola.1995.080331710

[23] Wodka TJ. Studies on the synthesis of block copolymers of acrylonitrile and ethylene oxide. Applied Polymer Science. 1993;**47**:407-416. DOI: 10.1002/app.1993.070470304

[24] Cakmak I. Preparation of multiphase block copolymers by redox polymerization process 2: Polymerization of acrylonitrile by the manganese(III)-poly(ethylene glycol) redox system. Die Angewandte Makromolekulare Chemie Banner. 1993;**211**:53-60. DOI: 10.1002/apmc.1995.052240105

[25] Erbil C. Blok copolymerization and chain extension by using PAAm prepolymer initiated with Ce(IV)-malonic acid. European Polymer Journal. 1999;**35**:1747-1754. DOI: 10.1016/S0014-3057(98)00282-1

[26] Sarac AS. Redox polymerization. Progress in Polymer Science. 1999;**24**:1149-1204

[27] Yagcı C, Yıldız U. Redox polymerization of methyl methacrylate with allyl alcohol 1,2-butoxylate-block-ethoxylate initiated by Ce(IV)/ HNO_3 redox system. European Polymer Journal. 2005;**41**:177-184. DOI: 10.1016/j.eurpolymj.2004.08.008

[28] Mıno G, Kaızerman S, Rasmussen E. The polymerization of acrylamide initiated by ceric nitrate-3-chloro-1-propanol redox systems. Journal of Polymer Science. 1959;**38**:393. DOI: 10.1002/pol.1959.1203813410

[29] Öz N, Akar A. Redox initiation system of ceric salt and dihydroxy poly(dimethylsiloxane)s for vinyl polymerization. Journal of Applied Polymer Science. 2006;**102**:2112-2116. DOI: 10.1002/app.24103

[30] Tunca U, Serhatli IE, Yagcı Y. Polymerization of acrylamide initiated by the redox system Ce(IV)-4,4′-azobis(4-cyano pentanol). Polymer Bulletin. 1989;**22**:483-488. DOI: 10.1007/BF00718923

[31] Arslan H, Hazer B. Ceric ion initiation of methyl methacrylate using polytetrahydrofuran diol and polycaprolactone diol. European Polymer Journal. 1999;**35**:1451-1455. DOI: 10.1016/S0014-3057(98)00221-3

[32] Arslan H, Eroğlu MS, Hazer B. Ceric ion initiation of methyl methacrylate from poly(glycidyl azide)-diol. European Polymer Journal. 2001;**37**:581-585. DOI: 10.1016/S0014-3057(00)00126-9

[33] Cakmak I, Hazer B, Yagci Y. Polymerization of acrylamide by the redox system ceric (IV) with poly(ethylene glycol) with azo groups. European Polymer Journal. 1991;**27**:101-103. DOI: 10.1016/0014-3057(91)90134-A

[34] Shimizu H, Yokohara T, Wada R, Okabe M. Influence of preparation conditions of poly (ethylene glycol)/ poly (N-isopropylacrylamide) block copolymers on their properties. Kōbunshi Rombun Shū. 2004;**61**: 640-642. DOI: 10.1295/koron.61.640

[35] Göktaş M. Synthesis and characterization of various block copolymers using PMMA-Br macroinitiator. Chemical Papers. 2019;**73**:2329-2339. DOI: 10.1007/s11696-019-00785-y

[36] Liu Y, Bai L, Zhang R, Li Y, Liu Y, Deng K. Block copolymerization of poly(ethylene glycol) and methyl acrylate using potassium diperiodatocuprate(III). Journal of Applied Polymer Science. 2005;**96**:2139-2145. DOI: 10.1002/app.21594

Section 2

Redox in Electrochemistry

Chapter 4

Redox Potentials as Reactivity Descriptors in Electrochemistry

José H. Zagal, Ingrid Ponce and Ruben Oñate

Abstract

A redox catalyst can be present in the solution phase or immobilized on the electrode surface. When the catalyst is present in the solution phase the process can proceed via inner- (with bond formation, chemical catalysis) or outer-sphere mechanisms (without bond formation, redox catalysis). For the latter, log k is linearly proportional to the redox potential of the catalysts, $E°$. In contrast, for inner-sphere catalyst, the values of k are much higher than those predicted by the redox potential of the catalyst. The behaviour of these catalysts when they are confined on the electrode surface is completely different. They all seem to work as inner-sphere catalysts where a crucial step is the formation of a bond between the active site and the target molecule. Plots of $(\log i)_E$ versus $E°$ give linear or volcano correlations. What is interesting in these volcano correlations is that the falling region corresponding to strong adsorption of intermediates to the active sites is not necessarily attributed to a gradual surface occupation of active sites by intermediates (Langmuir isotherm) but rather to a gradual decrease in the amount of M(II) active sites which are transformed into M(III)OH inactive sites due to the applied potential.

Keywords: redox potential, reactivity descriptors, redox catalysis, chemical catalysis, linear free-energy correlations, volcano correlations

1. Introduction

Predicting the rate of chemical processes on the basis of thermodynamic information is of fundamental importance in all areas of chemistry including biochemistry, coordination chemistry and especially electrochemistry [1]. Correlations do exist between the Gibbs free-energy for one series of reactions and logarithm of the reaction rate constant for a related series of reactions. These relations are known as linear free-energy relationships (LFER) [1]. For example, the Brønsted catalysis equation describes the relationship between the ionization constant of a series of catalysts and the reaction rate constant for a reaction on which the catalyst operates. The Hammett equation predicts the equilibrium constant or reaction rate constant of a reaction from a substituent constant and a reaction type constant. The Edwards equation relates the nucleophilic power to polarizability and basicity. The Marcus equation is an example of a quadratic free-energy relationship (QFER) that applies to electron transfer (ET) reactions where the activation energy is given by the inner and outer reorganizational energies. In the case of electrochemical reactions, the thermodynamics can be provided by the electrode potential or by the redox potential of a mediator or both [1].

The slope of the linear free-energy correlations reflects the sensitivity of rate constant to structural changes in a family of similar reactions. For organic reactions this slope resembles the definition of the parameter σ in Hammett correlations. It is interesting to mention that already in 1932 Frumkin stated [2] that at different electrode potentials, an electrochemical reaction is equivalent to a series of similar reactions differing only by the magnitude of $\Delta G°$, where $\Delta G°$ is equal to $-nF\Delta E°$ [3], n is the number of electrons exchanged in the complete reaction and F is the Faraday constant. In this case the role of substituents for changing the driving force is played by the potential of the electrode applied by an external source. The most common linear free-energy relationship in electrochemistry is the Tafel plot [4, 5], where ($\log i$) is plotted versus the potential of the electrode. This linear correlation is observed in the absence of mass transport limitations. If i is a kinetic current density, $\log i$ is proportional to $\log k$ (k is the rate constant), and the potential of the electrode is proportional to $\Delta G°$. As discussed further down, these correlations are not always linear in electron transfer processes in the homogenous phase or at electrode interfaces. In this chapter we will focus our attention of the correlations between the redox potential, a thermodynamic parameter of the catalyst and its catalytic activity. As we will discuss, these correlations are sometimes linear, and on other occasions they can have the shape of a volcano.

2. Catalytic effects in electrochemistry

A redox catalyst is a molecule that has an atom with two oxidation states that are kinetically more favorable than the oxidation states of the reactants and products upon which the catalyst is operating. In general, the most common active site is a metal that can be an atom that is coordinative unsaturated and is active for binding extraplanar ligands such as the reacting molecules [1]. The catalyst can be present in the solution phase or immobilized on the electrode surface. When the catalysts are present in the solution phase, the process can proceed via inner- or outer-sphere mechanisms [1]. For the latter, $\log k$ is linearly proportional to $E°$, whereas for inner-sphere catalyst, the values of k are much higher than those predicted by the formal potential of the catalyst. The behaviour of these catalysts when they are confined on the electrode surface is completely different. They all work as inner-sphere catalysts where a crucial step is the formation of a bond between the active site and the target molecule. Plots of $(\log i)_E$ versus $E°$ give volcano plots.

2.1 Redox catalysis and chemical catalysis

Metal, metal alloys and metal oxides have been studied extensively in the literature as catalysts for many reactions since the beginnings of electrochemistry. For example, the hydrogen evolution reaction (HER) was studied by Tafel [4], and he derived from his studies his well-known equation. In contrast, molecular catalysts have only been studied more intensively rather recently than the long history of fundamental and applied work using metals and alloys. It is important to point out that the electronic structure of metal electrodes is described using d-band theories where electronic levels form a continuum in the valence band. In contrast, molecular catalysts have discrete energy levels like any isolated chemical molecule or isolated atom. It is interesting then to correlate the electronic properties and energy levels of molecular catalysts with their catalytic activity for any reaction [6].

As mentioned above, a redox molecular catalyst is a species that can present at least two oxidation states that are kinetically much more favorable than the oxidation states of the reactants and products that the catalyst is promoting. Some electrochemical reactions involving reactants in the solution phase proceed at reasonable rates at potentials that are close to the equilibrium potential of the reaction, i.e. they require a rather low overpotential for the process to proceed at measurable rates. The kinetics of these reactions are then rather fast. A typical example is HER occurring on a Pt electrode [1]. However, many electrochemical reactions of interest usually require the transfer of more than one electron, each electron transfer step representing an energy barrier. The slow step, which is the rate-determining step, can be accelerated by the action of electrocatalysts mainly in two ways (some of them are illustrated in **Figure 1**):

i. The catalyst is in the solution phase, and the electrode serves only as a sink or source of electrons. The electrode regenerates the active form of the catalyst continuously which interacts with the target molecule in the homogeneous phase.

ii. The process is heterogeneous, and the catalyst can be the electrode surface itself (e.g. *d*-type noble metals like Pt, Pd, etc.) or a molecular catalyst confined on a rather inert electrode surface (i.e. for so-called modified electrodes). In the case of a molecular catalyst attached to the electrode surface, the electrode also regenerates the active form of the catalyst continuously.

2.1.1 Case (i): the catalyst is present in the solution phase

In case (i), the homogeneous catalytic process can proceed via two different pathways: via outer-sphere and inner-sphere processes [7].

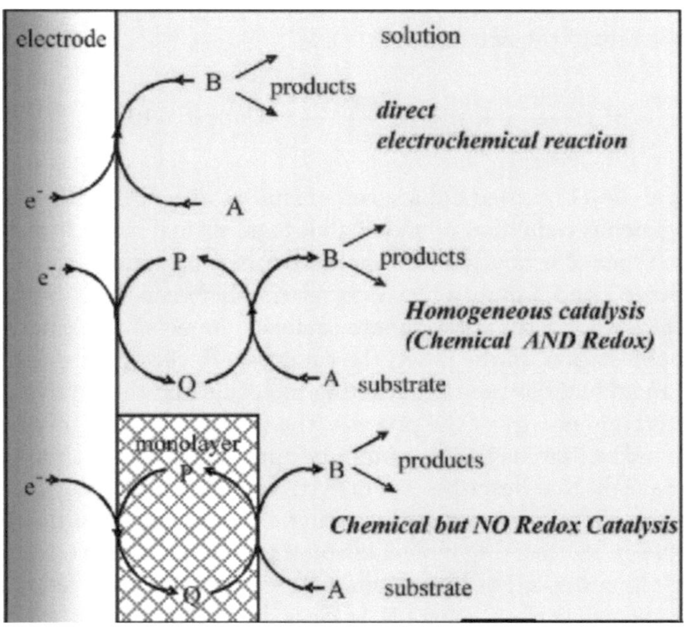

Figure 1.
Different reactions schemes for catalytic processes in electrochemistry as described by Savéant [7] (reproduced by permission John Wiley & son).

(ia) For the case of an outer-sphere reaction, the process has been termed “redox catalysis”, and the active form of the catalysts is present in a given oxidation state. The catalyst oxidation state changes upon interacting with the target molecule, and the active state is continuously regenerated at the electrode surface. The catalyst and the reactant only collide in the homogeneous phase without forming a bond. The catalyst recuperates its initial oxidation state at the electrode surface as illustrated in **Figure 1**. The catalysis in this case lowers the overpotential of the overall reaction by acting in a three-dimensional dispersion. The redox potential of the catalyst contributes to the driving force of the reaction, not the potential of the electrode, which only serves to regenerate the active form of the catalyst at the interface (no reaction takes place directly between the electrode and the reactants [7]).

(ib) For the case of an “inner sphere” process, the process has been termed “chemical catalysis” and involves the temporary formation of an adduct between the mediator and the reactant. The bond formed between the reactant and the catalyst is broken after the exchange of electrons to form intermediates and products. This regenerates the catalysts that recuperate its initial oxidation state at the electrode [7].

If we take as an example a reduction reaction mediated by catalysts dissolved in the solution phase, on thermodynamic grounds one would expect that the more negative the formal potential of the mediator (more powerful reductant), the higher its reactivity for the oxidation of the target, according to the reaction scheme below, where step in Eq. (3) is rate-determining and the first step, Eq. (1), is the electrogeneration of the catalyst in its active form:

$$[non - active\ OX] + e^- \rightarrow [active\ RED]^- \tag{1}$$

$$[active\ RED]^- + target \rightarrow [non - active\ OX] + target^- \tag{2}$$

$$target^- \xrightarrow{k} products \tag{3}$$

According to this reaction scheme and assuming that step 3, Eq. (3), is rate controlling, the rate of the reaction is:

$$\frac{-d[target]}{dt} = \frac{k\mathrm{K}[target][non - activeOX]}{[activeRED]^-} = k[target] * 10^{(E-E^\circ)/0.059} \tag{4}$$

Savéant et al. [7–11] carried out a series of studies using different redox catalysts for the homogeneous reduction of alkyl halides and vicinal halides to alkenes. They identified two types of catalytic systems: outer-sphere and inner-sphere redox catalysts. **Figures 2** and **3** show a linear correlation between the E° formal potential of the catalysts and log k for outer-sphere catalysts. Inner-sphere catalysts fall out for the linear correlation. In the latter, the catalytic effect originates from the formation of an adduct between the reacting molecule and the catalyst, which lowers the activation energy of the process. The reaction studied involved the rupture of a bond so Marcus theory cannot be applied but rather a modified version proposed by Savéant that describes a Morse curve rather that a parabola for the reaction products [11]. In the latter, the catalytic effect rises from the formation of an adduct between the reacting molecule and the catalyst, which lowers the activation energy of the process. For inner-sphere processes where the redox catalysts are in the homogeneous phase, there is no clear dependence of the rate of the process of the redox potential of the catalyst. The inner-sphere process has lower activation energy than the outer-sphere pathway. If both pathways are available, the inner-sphere process is preferred.

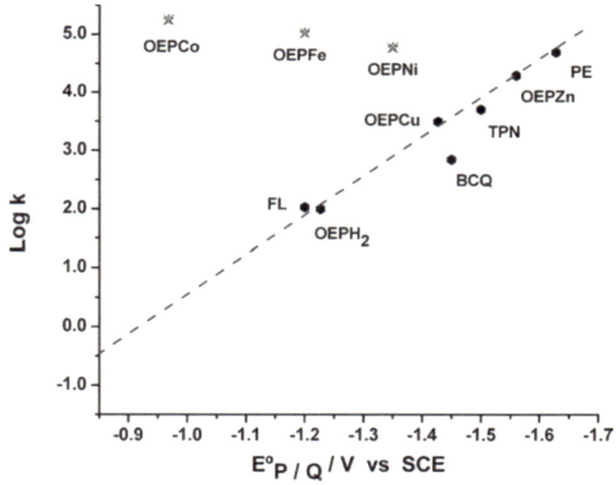

Figure 2.
Dependence of log k *versus the redox potential of the catalyst for the reduction of trans-1,2 dibromocyclohexane by aromatic radicals and by reduced metalloporphyrins from Lexa et al. [11] (reproduced by permission of the American Chemical Society).*

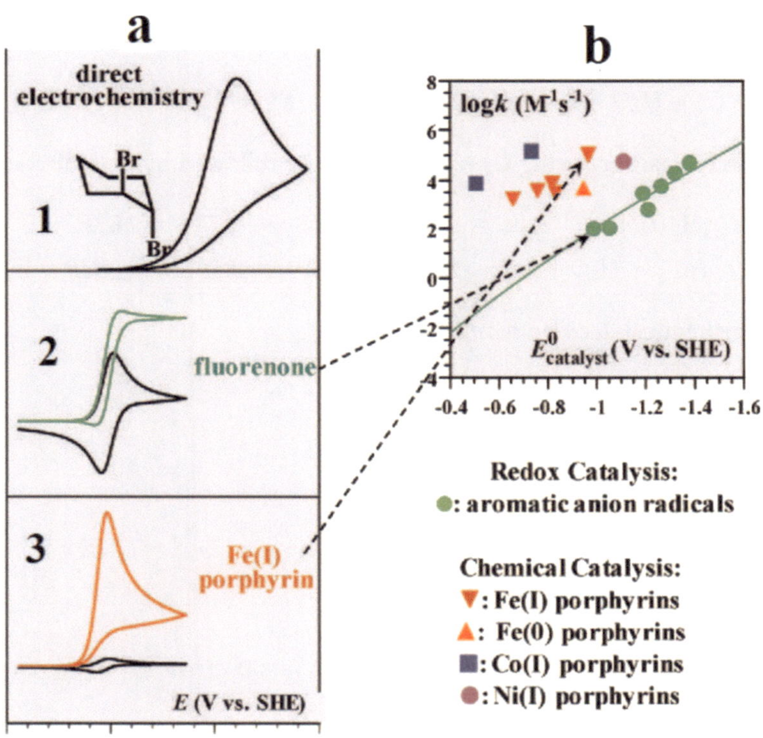

Figure 3.
Homogeneous redox and chemical catalysis of trans-1,2 dibromocyclohexane. (a1) Cyclic voltammogram in DMF of direct electrochemical reduction on glassy carbon, (a2) redox catalysis of fluorenone and (a3) chemical catalysis promoted by iron(I) porphyrin. (b) Rate constant versus the standard redox potential of the catalyst (taken from Savéant [7]. Reproduced by permission John Wiley & son).

2.1.2 Electrocatalysis by redox mediators, chemical catalysis and oxygen reduction

Electrodes modified with macrocyclic metal complexes that exhibit fast reversible redox processes centred on the metal have received considerable attention in the literature because of the applications in electrocatalysts and also as electrochemical sensors [12–14]. A very important area that has been investigated by many authors for decades is the oxygen reduction reaction (ORR). The ORR is a very important reaction indeed as it is involved in energy conversion processes (fuel cells, air batteries) in metallic corrosion etc., without considering important biological processes as the respiration chain [15]. The complete reduction of O_2 in aqueous media requires the presence of catalysts and involves the transfer of four electrons and the splitting of the O–O bond. This reaction delivers the most energy in a fuel cell and in living systems. However, in electrochemistry rather few electrode materials promote the four-electron reduction of O_2.

The ORR process is a multielectron reaction in aqueous media that occurs via two main pathways: one involving two electrons to give peroxide and the direct four-electron pathway to give water. The four-electron process involves the rupture of the O–O bond [16]. The nature of the electrode surface strongly influences the preferred pathway. Most electrode materials catalyse the reaction only via two electrons to form peroxide:

Pathways for ORR.

In acid media:

$$O_2 + 2H^+ + 2e^- \rightarrow H_2O_2 \qquad E^o = +0.67\ V\ vs\ NHE \tag{5}$$

In alkaline media:

$$O_2 + H_2O + 2e^- \rightarrow HO_2^- + OH^- \qquad E^o = -0.65\ V\ vs\ NHE \tag{6}$$

Peroxide formation during O_2 reduction can be followed by its reduction:

$$H_2O_2 + 2H^+ + 2e^- \rightarrow 2H_2O \qquad E^o = +1.77\ V\ vs\ NHE \tag{7}$$

$$HO_2^- + H_2O + 2e^- \rightarrow 3OH^- \qquad E^o = +0.867\ V\ vs\ NHE \tag{8}$$

or by its chemical decomposition:

$$2H_2O_2 \rightarrow 2H_2O + O_2 \tag{9}$$

Direct four-electron reduction pathway.

In acid:

$$O_2 + 4H^+ + 4e^- \rightarrow 2H_2O_2 \qquad E^o = +1.229\ V\ vs\ NHE \tag{10}$$

In alkaline:

$$O_2 + 2H_2O + 4e^- \rightarrow 4OH^- \qquad E^o = +0.401\ V\ vs\ NHE \tag{11}$$

In strongly alkaline solutions or in organic solvents, O_2 is reduced via the transfer of a single electron to give superoxide ion. This process is outer-sphere:

$$O_2 + e^- \rightarrow O_2^- \qquad E^o = -0.33\ V\ vs\ NHE \tag{12}$$

The O_2 four-electron reduction reaction is thermodynamically spontaneous in O_2/H_2 fuel cells, but its kinetics are slow on most electrode materials. The sluggishness of the reaction kinetics can be attributed to the transfer of four electrons involving the formation of bonds between intermediates and the active sites. The best catalytic materials for the four-electron ORR contain Platinum. The high cost of this metal is one of the limitations for the widespread use of fuel cells. Several authors have faced the challenge of preparing catalytic materials for oxygen reduction that do not involve precious metals like Platinum and its alloys. This probably started with the seminal work of Jasinski [17] that reported that cobalt phthalocyanine (CoPc) exhibited catalytic activity for the ORR. This discovery triggered the research on metal macrocyclic MN4 complexes as potential catalysts for ORR. **Figure 4** shows the molecular structures of the most common macrocyclic complexes investigated as catalysts for ORR. However, some MN4 electrodes containing these complexes are not stable in the corrosive environment of a fuel cell after long operating conditions. For this reason, a new research started, on pyrolysed MN4 complexes [18]. Heat treatment at different temperatures increases both activity and stability, and this has been demonstrated by several authors. It was also found that MN4 or MNx pyrolysed catalytic materials can be obtained without using MN4 complexes as a starting ingredient, but rather using metal salts and N-containing organic compounds.

In this chapter, we will discuss the different reactivity predictors that serve as guidelines for the synthesis of more active catalysts [6, 12, 14, 16, 19–22] but with emphasis on the redox potential of the catalyst. There are several reactivity predictors for MN4 macrocyclic complexes that have been described in the literature: (i) the *d*-level populations in the central metal, (ii) the donor-acceptor intermolecular hardness, (iii) the M-O_2 binding energy and (iv) the $M^{+n}/M^{+(n-1)}$ redox potential. We will discuss predictors (i) and (iii)–(iv), especially the last two, the M-O_2 binding energy and the redox potential of the catalysts, as they seem to be related to each other. One rather simple reactivity predictor is the number of *d*-electrons when comparing the activities of a family of similar macrocyclic metal complexes. For example, when comparing the activities of metal phthalocyanines (MPcs), a plot of E at constant current versus the number of *d*-electrons gives a

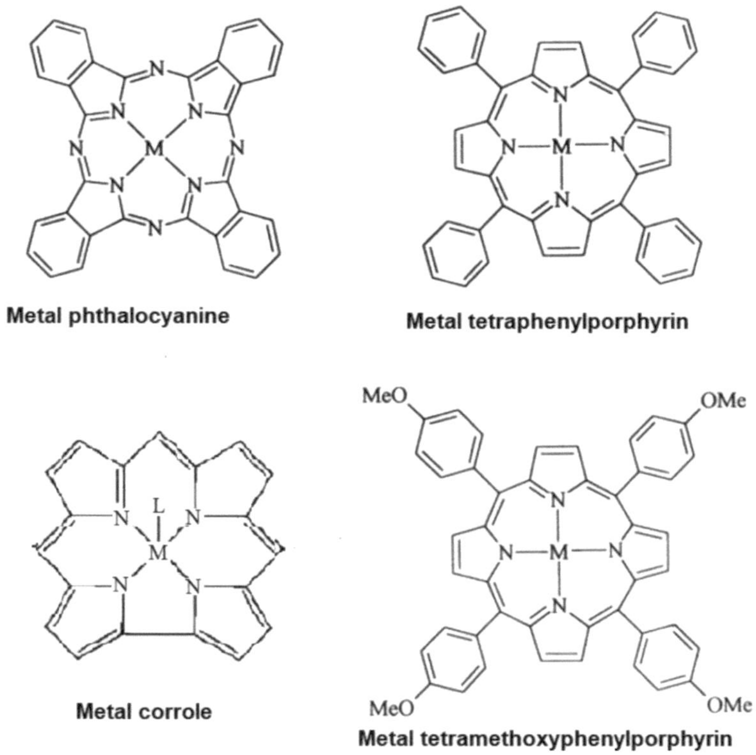

Figure 4.
Structure of several unsubstituted and substituted metal phthalocyanines, metal porphyrins and a metal corrole.

parabolic correlation with FePc showing the highest activity with 6 *d*-electrons. When metal porphyrins (MPs) are compared, the highest activity is exhibited by CoPs with 7 *d*-electrons. These results are illustrated in **Figure 5**. These results indicate that highest activity is achieved by complexes with nearly half-filled *d*-orbitals. However, it is important to point out that those metals with low and or partially filled *d*-orbitals exhibit reversible redox processes located on the metal centre. This is true for Cr, Mn, Fe and Co macrocyclic complexes. The frontier orbitals of these complexes have *d*-character and have the proper symmetry to bind an extraplanar ligand like O_2. On the other hand, Ni, Cu and Zn complexes exhibit no redox process centred on the metal and redox processes involve the ligand. These complexes show less catalytic activity for ORR and other reactions. We will discuss further down the important role that the redox potential plays in dictating the catalytic activity. As a graphical example, **Figure 6** illustrates electron tunneling microscopy images obtained with CoPc and CuPc by two different authors [12, 19]. It is clear that CoPc shows an electron-rich zone on the central metal in contrast to CuPc. This is corroborated in the same **Figure 6C** by frontier orbital profiles

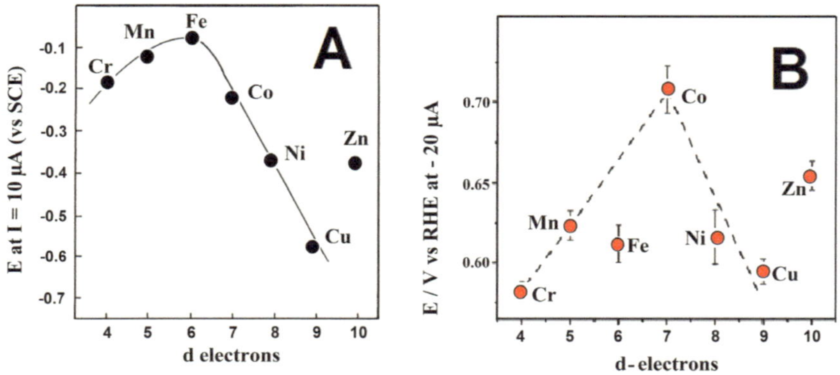

Figure 5.
Correlation between electrocatalytic activity (as potential at constant current) versus the number of d*-electrons for ORR in alkaline media on metallo-phthalocyanines (A) and metal porphyrins (B) (data on MPcs from [12] and MPs from [19]).*

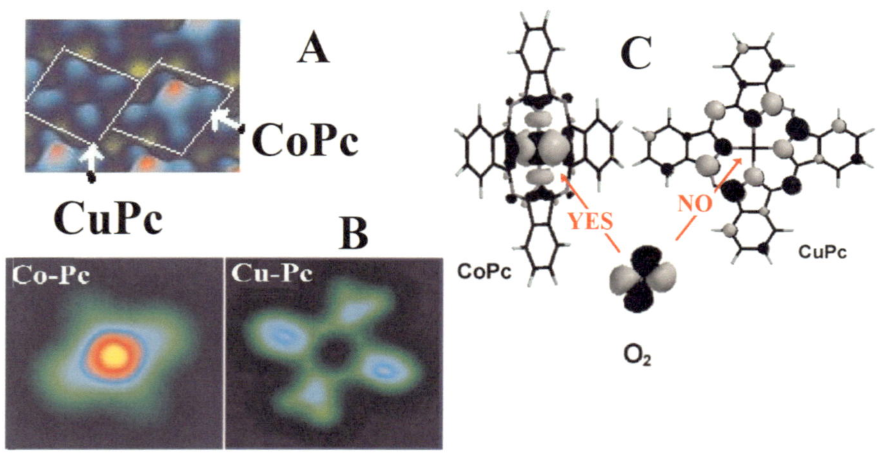

Figure 6.
(A and B) tunnel microscopy images of CuPc (inactive) and CoPc (active) adsorbed on Au (adapted from [20, 21]). (C) theoretical profiles of frontier orbitals of CoPc and CuPc (taken from [22] with permission from Elsevier).

obtained theoretically that show that CoPc exhibits a prominent d-orbital that can bind O_2 in contrast again to CuPc. These images are in agreement with the reactivity trends shown in **Figure 5A**.

Figure 7 illustrates a correlation of catalytic activity as $(\log i)_E$ for ORR at constant potential versus the M(III)/(II) redox potential of the catalyst; the data was obtained in acid solution (8 M H_2SO_4) [23]. It is clear from this figure that the activity increases almost linearly with the redox potential of the catalyst up to a point and then decreases. A maximum activity is observed for octaethyl CoPs (see CoOEP in **Figure 7**). It is interesting that Fe complexes appear on one side of the volcano and Co complexes on the other side. These volcano-shaped correlations are very common in electrocatalysis and heterogeneous catalysis, but instead of the redox potential a parameter is used that describes the degree of interaction of O_2 with the active site. In recent work the M-O_2 binding energy is used [6], and this parameter is the most common reactivity descriptor. The data in **Figure 7** strongly suggests that the redox potential of the catalyst is a reactivity descriptor and then should be related to the M-O_2 energy.

Figure 8 clearly shows that there is a direct correlation between the M(III)/(II) redox potential of the catalyst and the M-O_2 binding energy, for redox data obtained in alkaline media [6]. The data in **Figure 8** indicates that the M-O_2 binding is a reactivity descriptor for ORR catalysed by MN4 macrocyclic complexes as is well established for metal electrodes [24]. Strong binding catalysts appear on the left side of the volcano, and weak binding catalysts appear on the right-hand side of the correlation, illustrating that these catalysts follow the Sabatier principle that for achieving the highest catalytic activity, the binding needs to be not too strong and not too weak. The volcano correlation illustrated in **Figure 8** bears a strong resemblance with similar volcano correlations obtained for ORR catalysed by pure metals [24]. For both MN4 complexes and pure metals, strong binding catalysts promote the four-electron reduction of O_2 to water or OH^-, whereas weak binding catalysts only promote the two-electron reduction to hydrogen peroxide.

Figure 9 illustrates the potentiodynamic response of CoPc/OPG and FePc/OPG, where OPG is ordinary pyrolytic graphite electrode, and also the reduction wave for

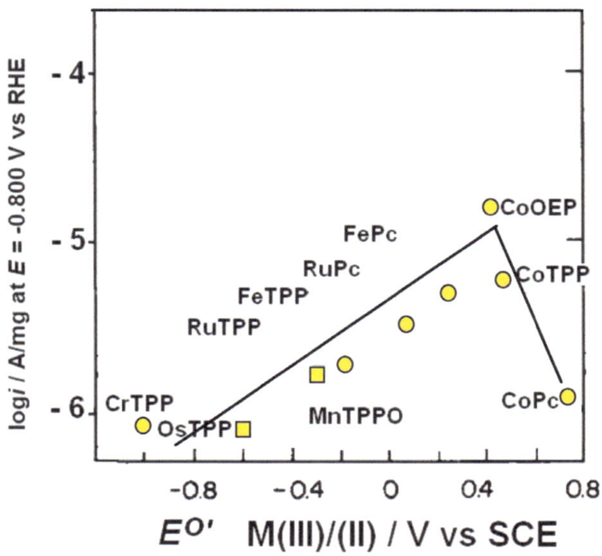

Figure 7.
Illustration of volcano correlations of catalytic activity as (log i)$_E$ for ORR at constant potential versus the M (III)/(II) redox potential of the catalyst. The data was obtained in acid solution (8 M H_2S_4) [23].

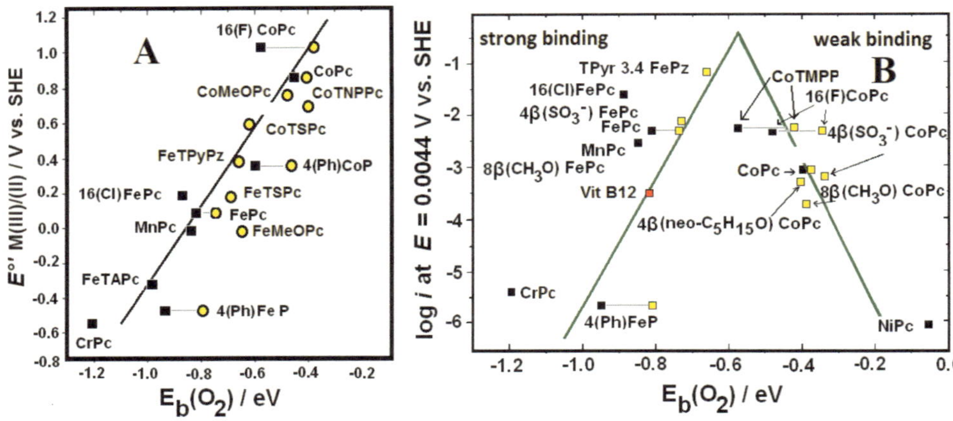

Figure 8.
(A) Correlation of electrocatalytic versus M(III)/(II) redox potential of the MN4 complexes and (B) versus the M-O_2 binding energy to the metal center in the MN4 complex (ref.[6] and references there in).

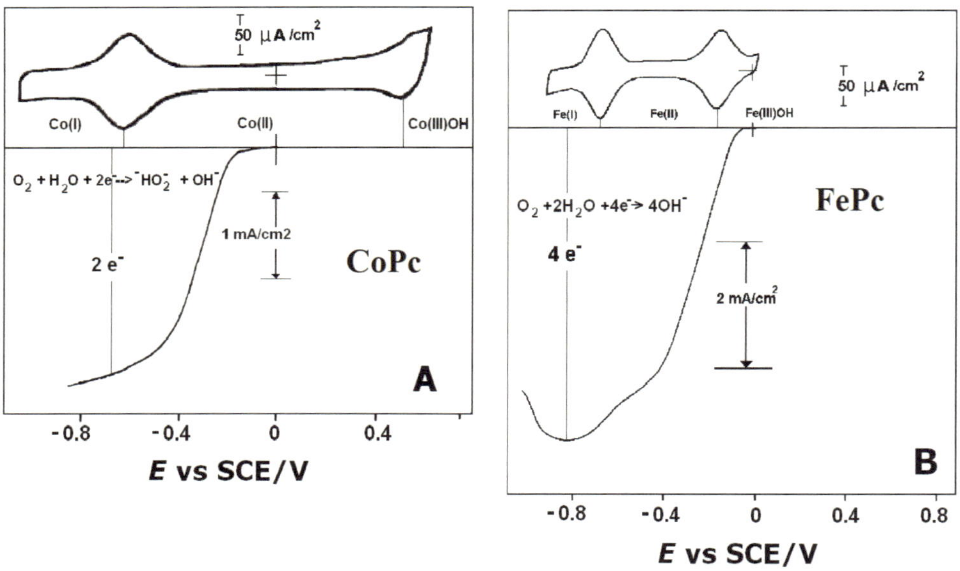

Figure 9.
(A) Cyclic voltammogram obtained in 0.1 M NaOH with an ordinary pyrolytic graphite (OPG) disk electrode modified with a monolayer of CoPc in the absence of O_2. The bottom of that figure shows a polarization curve obtained with the same electrode, in the presence of O_2 (saturated) and with rotation of 1000 rpm. (B) Analogous to (A) but with the OPG electrode modified with FePc [25].

ORR obtained on rotating OPG disk modified with monolayers of CoPc or FePc. Both complexes exhibit M(II)/(I) and M(III)/(II) reversible redox couples in the potential range examined. However, there are some striking differences in the response of these two complexes: first, the Co(II)/(I) and Co(III)/(II) redox couples appear much more separated for CoPc than similar processes occurring on FePc. For CoPc the onset for ORR appears at potentials between the Co(II)/(I) and Co(III)/(II) transitions, and the onset is well removed in the positive direction from the Co (III)/(II) redox potential, whereas for FePc, the onset for ORR appears at a potential close and more positive than the Fe(III)/(II). In this case ORR starts at potentials where the surface coverage by Fe(II) active sites is potential-dependent, and this coverage gradually increases as the potential is scanned to more negative potentials [25]. **Figure 10** illustrates the potential dependence of the coverage of the surface by

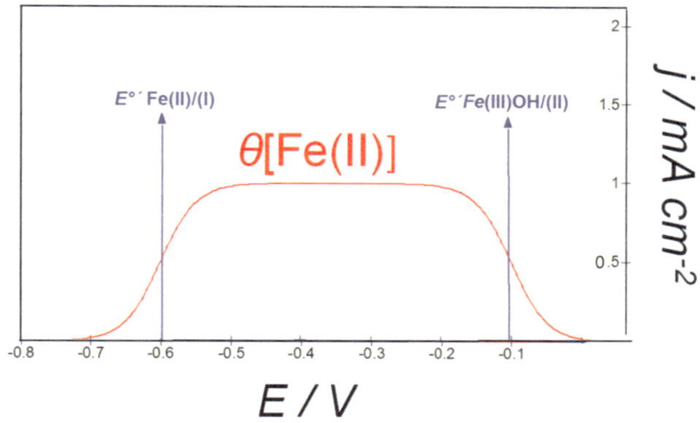

Figure 10.
Mathematical simulation of the variation of fractional coverage $\theta_{Fe(II)}$ on the electrode surface as a function of potential for two hypothetical values of $E_{Fe(II)/(I)}$ and $E_{Fe(III)/(II)}$. Simulation involved the Nernst equation applied to surface confined species, assuming ideal behaviour (adapted from [26]).

FePc in the oxidation state (II) simulated using Eq. (13). This profile is obtained by simply using the Nernst equation for adsorbed species. This figure can explain the decrease in catalytic activity at potentials below the $E^{\circ}_{Fe(II)/(I)}$ transition as it appears that Fe(I) is not such a good catalyst as Fe(II) and the selectivity of the reaction changes from four electrons to two electrons to give peroxide. That could explain the decline in the ORR currents at more negative potentials. For simplicity in Eq. (13), the concentration of OH^- was taken as unity:

$$\theta(Fe(II)) = \frac{exp^{\left(F/RT\left(E-E^{\circ}_{II/I}\right)\right)}}{\left[1+exp^{\left(F/RT\left(E-E^{\circ}_{II/I}\right)\right)}\right]} \frac{1}{\left[1+exp^{\left(F/RT\left(E-E^{\circ}_{III/II}\right)\right)}\right]} \quad (13)$$

2.1.2.1 *Effect of redox potential on the electrocatalytic activity of MN4 complexes for ORR*

The catalytic properties of MN4 macrocyclics for ORR have been reviewed by many authors [1, 6, 12–16, 18, 19, 23, 25, 27, 28]. Catalytic activity and stability can also be improved by heat treatment of these complexes combined on carbon. The amount of literature on this topic is very large and is beyond the scope of this manuscript to discuss it in any detail. As pointed out above, there are several reviews on this subject [6, 18, 23, 29–31], so discussion will be focused on using the complex redox potential as a reactivity descriptor.

Randin was the first to attempt to rationalize the catalytic activity of MN4 complexes [32] in 1974 later followed by Beck in 1977 [33] using simply the redox potential of the MN4 complex as a reactivity indicator. A first theoretical explanation based on these terms was provided by Ulstrup in 1977 [34]. According to Randin and Beck, during the binding of O_2 to the metal ion in the complex, the metal is partially oxidized, thereby reducing the O_2 molecule according to the:

$$M(II) + O_2 \rightarrow \left[M(III)O_2^-\right] \quad (14)$$

$$\left[M(III)O_2^-\right] \rightarrow products + M(III) \quad (15)$$

In order to account for supplementary experimental evidence, a somewhat modified model was proposed by Beck in which the central metal ion might also be only partially oxidized. According to reactions in Eqs. (14) and (15), the potential at which O_2 is reduced should be closely related to the M(III)/M(II) redox potential of the central metal ion. Later studies published by van Veen et al. [23] using measurements of the redox potential of the complex under the same conditions as those under which the ORR kinetic data were obtained showed for the first time that the activity of several metal complexes, plotted as log i at constant potential versus the M(III)/(II) redox potential of the catalyst, gave a volcano-shaped curve (**Figure 7**).

More recent results comparing the activity of a large collection of MN4 catalysts for ORR in alkaline media is shown in **Figure 8**. The data was reported by Zagal and Koper [6]. The volcano correlation illustrated in **Figure 8** shows two features. On the weak binding side, the two-electron reduction catalysts appear. According to the Sabatier principle for maximum catalytic activity, the binding of the reacting molecule needs to be not too strong and not too weak and to essentially explain the shape of the correlation where low activity is observed for low M-O_2 binding and for high M-O_2 binding energies. The highest activity is achieved for intermediate M-O_2 binding energies. The shape of the volcano correlation can also be explained using a Langmuir isotherm that essentially defines the coverage (θ) of M-O_2 species on the catalytic surface. Catalysts binding O_2 very weakly give very low values of θ and low activities, and the opposite is true for catalysts binding O_2 very strongly which give values of θ approaching unity, and in that case practically all the surface is occupied with adsorbed intermediates, blocking the active sites and again giving very low activities. These two extreme cases represent the foothills of both sides of the volcano correlation. It can be demonstrated that under standard conditions, the apex of the volcano corresponds to a situation where $\theta = 0.5$ and $\Delta G^{o}_{O2} = 0$. For simplicity we consider the binding of O_2 to the active sites occurring with the transfer of a first electron, Eq. (17). This is generally the bottleneck of the whole reduction process leading to peroxide (two-electron transfers) or to water (four-electron transfer):

$$[M(III)OHN_4]_{ad} + e^- \rightarrow [M(II)N_4]_{ad} + OH^- \tag{16}$$

$$[M(II)N_4]_{ad} + O_2 + e^- \rightarrow [O_2 - MN_4^-]_{ad} \tag{17}$$

Depending on the catalyst, this equation will be shifted to the product (strong adsorption) or to the reactants (weak adsorption). Other adsorption steps can occur, involving other intermediates. The kinetic current density for a given potential consistent with the Butler-Volmer equation is given by the following equation neglecting the back reaction:

$$i = nFk\, \theta_{M(II)}\, p_{O2}\, (1-\theta)\ exp^{-\beta FE/RT}\ exp^{-\beta' \Delta G^{o}_{ad}/RT} \tag{18}$$

where β is the symmetry factor of the energy barrier and β' the is the Brönsted-Polanyi coefficient. β and β' can be assumed to be 0.5.

From the data in **Figure 8** we can assume that $\Delta G^{o}_{ad} = nFE^{o} + C$, where E^{o} = is the M(III)/(II) formal potential of each complex. The surface coverage of active sites varies as $0 < \theta_{M(II)} < 1$ depending on the electrode potential. **Figure 10** shows a simulation of the variation of $\theta_{M(II)}$ with potential using the Nernst equation for adsorbed species [26].

However, in the case of metal complexes with very negative redox potentials, the Fe(III)/Fe(II) redox transition also comes into play, lowering the fraction of catalytically active sites θ[M(II)] as illustrated in **Figure 10**. The M(II) active sites will predominate in the potential range, $E^{o'}_{(II/I)} \leq E \leq E^{o'}_{(III/II)}$, as illustrated in **Figure 10**.

Since the Fe(III) binds OH^- [35], these sites will be inactive for the reduction of O_2; the coverage of adsorbed O_2 can be assumed to follow a Langmuir isotherm:

$$\theta = \frac{p_{O2}\ exp^{(-\Delta G^{\circ}_{O2}/RT)}}{\left[1 + p_{O2}\ exp^{(-\Delta G^{\circ}_{O2}/RT)}\right]} \tag{19}$$

In volcano plots, the data for different catalysts is compared at constant E, so the Butler-Volmer exponential term for simplicity can be absorbed into the constant k'. For small coverages, ΔG°_{O2} is positive so Eq. (19) can be written as:

$$k' = k\ exp^{(-\beta FE/RT)} \tag{20}$$

$$i = nFk'\theta_{M(II)}\, p_{O2}\, (1 - \theta_{ad})\ exp^{(-\beta' \Delta G^{\circ}_{ad}/RT)} \tag{21}$$

$$(1-\theta) = \frac{1 + p_{O2}\ exp^{(-\Delta G^{\circ}_{O2}/RT)}}{\left[1 + p_{O2}\ exp^{(-\Delta G^{\circ}_{O2}/RT)}\right]} - \frac{p_{O2}\, exp^{(-\Delta G^{\circ}_{O2}/RT)}}{1 + p_{O2}\, exp^{(-\Delta G^{\circ}_{O2}/RT)}} \tag{22}$$

$$(1-\theta) = \frac{1}{\left[1 + p_{O2}\ exp^{(-\Delta G^{\circ}_{O2}/RT)}\right]} \tag{23}$$

$$i = \theta_{Fe(II)} \frac{nFk' p_{O2}\ exp^{(-\beta' \Delta G^{\circ}_{O2}/RT)}}{\left[1 + p_{O2}\ exp^{(-\Delta G^{\circ}_{O2}/RT)}\right]} \tag{24}$$

Equation (24) essentially describes the shape of the volcano plot of log i vs. ΔG°_{O2} and that the maximum current density will be observed for $\Delta G^{\circ}_{O2} = 0$, and the maximum current at the apex of the volcano is:

$$i = \frac{nFk'\theta_{Fe(II)}\, p_{O2}}{\left[1 + p_{O2}\right]} \tag{25}$$

For strong adsorption, ΔG°_{O2} is large with a negative sign so $1 << p_{O2}exp\ (-\Delta G_{O2}/RT)$, so the general Eq. (25) becomes:

Strong adsorption:

$$i = nFk'\ \Gamma\theta_{Fe(II)}\ exp^{(+\beta' \Delta G^{\circ}_{O2}/RT)} \tag{26}$$

Equation (26) explains the linear correlation in the region of strong adsorption (ΔG°_{O2} is negative) and is essentially independent of the concentration of the reactant, in this case, O_2.

For weak adsorption, ΔG°_{O2} is very positive, the term $pO_2exp(-\Delta G^{\circ}_{O2}/RT) << 1$ in Eq. (24) vanishes, and the general Eq. (24) becomes simply:

Weak adsorption:

$$i = nFk'\ \theta_{Fe(II)}\, p_{O2}\ exp^{(-\beta' \Delta G^{\circ}_{ad}/RT)} \tag{27}$$

So, Eqs. (26) and (27) describe both linear correlations in the volcano plots with slopes of different signs, i.e. $+\beta'/RT$ for the strong adsorption region and $-\beta' \Delta G^{\circ}_{O2}/RT$ for the weak adsorption region. According to this, the volcano plot should be symmetrical since both legs of the volcano have the same absolute value of the slope. It is necessary to clarify that in theoretical calculations, the binding energy ΔE_{bO2} is used because it does not contain entropy terms as ΔG°_{O2} and the entropic terms are difficult to estimate. The binding energy is essentially the energy to break the M-O_2 bond. If we use $\Delta G^{\circ}_{ad} = nFE^{\circ} + C$ and replacing the formula described above, the constant "C" can be adsorbed in a new constant k":

Strong adsorption:

$$i = nFk''\theta_{M(II)}\ exp^{\left(+\beta' E^{\circ}_{Fe(III)(II)}F/RT\right)} \tag{28}$$

Weak adsorption:

$$i = nFk''\theta_{M(II)}\, p_{O2}\ exp^{\left(-\beta' E^{\circ}_{Fe(III)(II)}F/RT\right)} \tag{29}$$

It is important to point out that β is the symmetry factor of the electron transfer energy barrier and β' is a Brönsted-Polanyi factor that reflects the effect of the adsorption on the activation energy. Both Eqs. (28) and (29) predict that the currents increase from positive $E°$ (weak adsorption) to negative (strong adsorption) values of $E°$ but up to 1 point. The highest activity will be at the intersection of those two lines. As the fraction of adsorbed intermediates θ increases, the currents start to decrease. This originates the "falling" side of the volcano, and the term (1-θ) tends to zero as $\theta \to 1$. This corresponds to the left side of the volcano correlation in **Figure 8**.

There is an alternative and more realistic explanation for the falling of the currents in the so-called "strong adsorption region". Complexes located in that particular region are not necessarily in the M(II) active state. These complexes present Fe(III)/(II) redox potentials that are more negative than the electrode potentials used for comparing the activity. In those cases $\theta_{Fe(II)}$ can be lower than 1.

If we consider just the Fe(III)/(II) couple, as E is well above the $E^{o}{}_{\mathrm{Fe(II)/(I)}}$ redox potential, we can use the Nernst equation for adsorbed species:

$$\theta Fe(II) = \frac{1}{1 + exp^{(F(E-E^{\circ})/RT)}} \tag{30}$$

where, for simplicity $E^{o} = E^{o}{}_{\mathrm{Fe(III)/(II)}}$, $\theta_{Fe(II)}$ will become gradually smaller as $E >> E^{o}{}_{Fe(III)/(II)}$. Eq. 30 carries the assumption that the adsorbed MN4 species behave ideally.

So if we consider that the falling region of the volcano correlation is attributed primarily to a gradual decrease in $\theta_{Fe(II)}$ and not to a gradual decrease in $(1 - \theta)$, the amount of active sites is not then blocked by adsorbed intermediates. We can then introduce $\theta_{Fe(II)}$ *in Eq. (31):*

$$i = nFk''\theta_{M(II)}\, p_{O2}\ exp^{(-\beta' E^{\circ}F/RT)} \tag{31}$$

$$i = \frac{nFk''M(II)p_{O2}\, exp^{(-\beta' E^{\circ}F/RT)}}{1 + exp^{(F(E-E^{\circ})/RT)}} \tag{32}$$

For those catalysts that have formal potentials below the electrode potential, $E° << E\ expF/RT(E - E°) >> 1$, then the equation becomes:

$$i = nFk''p_{O2}\, exp^{[(-\beta' E^{\circ}F/RT)-((EF/RT)-(E^{\circ}F/RT))]} \tag{33}$$

$$i = nFk''p_{O2}\, exp^{[(+\beta'(E^{\circ}-E)F)/RT]} \tag{34}$$

Equation (34) predicts that in the falling region of the volcano correlation, hypothetically the region of strong adsorption, the plot of $(\log i)_E$ versus $E°$ should be linear with a slope of $+RT/\beta' F$. Assuming $\beta' = 0.5$ then the slope should be +0.120 V/decade which is close to the experimental value of +0.140 V/decade.

The gradual decrease in the currents on the falling region can then be attributed to the gradual decrease in $\theta_{Fe(II)}$ as the potential becomes more negative which somehow resembles what happens on metal electrodes where the surface becomes gradually covered with adsorbed intermediates. The volcano correlation where currents have been corrected for $\theta_{Fe(II)}$ becomes a linear correlation where the activity as log i increases with the driving force of the catalyst (see linear correlation in **Figure 11**). The slope is close to -0.120 V/decade again, very similar to the experimental value

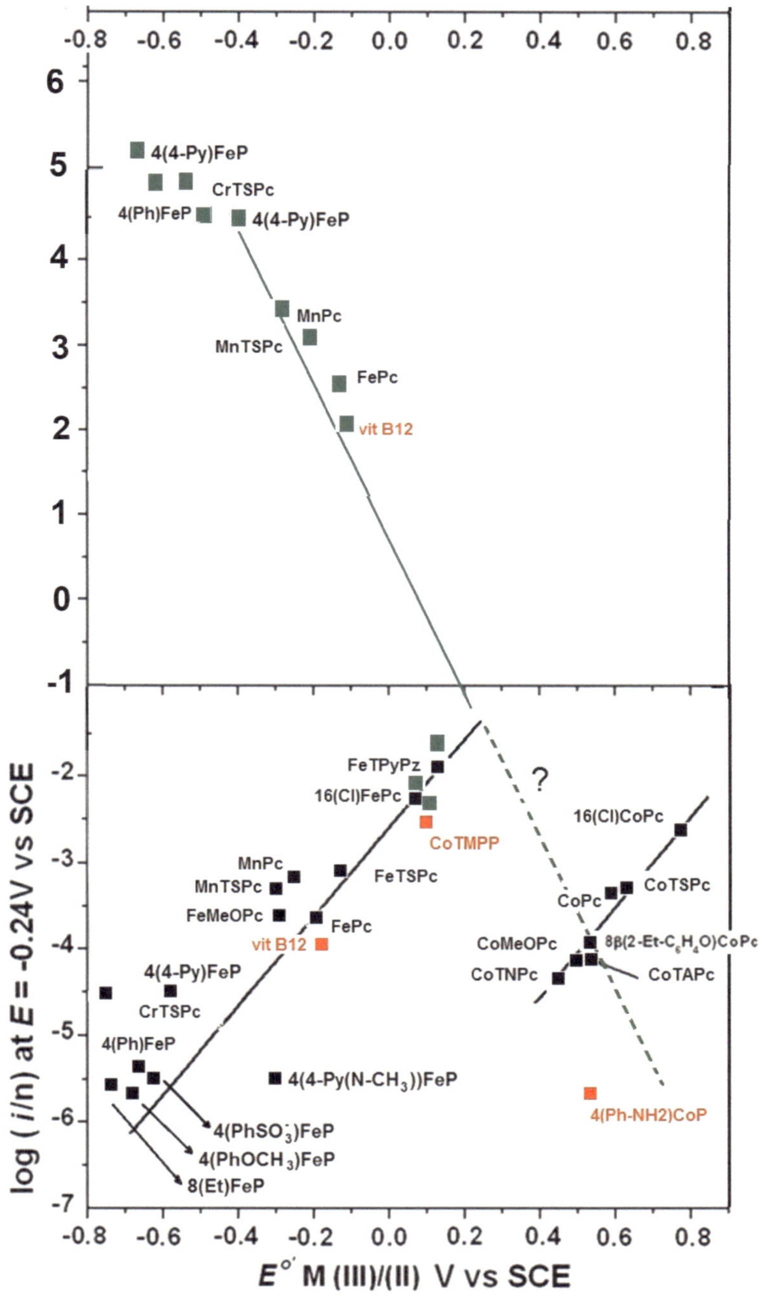

Figure 11.
Volcano correlation of $\log(i/n)_E$ *(*$n = 4$ *for Mn and Fe complexes and* $n = 2$ *for Co complexes, except vitamin B12) versus* $E_{M(III)/(II)}$ *for ORR. The linear correlation involves currents corrected for* $\theta_{M(II)}$ *as* $\log(i/\theta_{M(II)}\, n)_E$*. The only complexes affected are those having* $E_{M(III)/(II)} << E$ *(adapted from [6]).*

of −0.130 V/decade. This also suggests that the redox potential of the catalysts acts as the driving force of the reaction, in a similar way of the electrode potential. The slope of this correlation is a linear free-energy correlation similar to a Tafel line where the driving force is provided by the overpotential. In inner-sphere chemical catalysis or electrocatalysis, the reacting molecule, in this case O_2, does react promoted by two independent parameters: the electrode potential and the redox potential of the catalyst. The redox potential of the catalyst represents an excellent reactivity descriptor as it can be measured under the same conditions of the kinetic experiments. It is important to point out that the reactivity guidelines provided by MN4 macrocyclic complexes are also valid for smaller complexes like Cu phenanthrolines where the activity also depends in this case on the Cu(II)/(I) redox potential where the active state is Cu(I). For these complexes the activity of $(\log i)_E$ increases with the Cu(II)/(I) redox potential according to a linear correlation with a slope close to +0.120 V/decade. No volcano correlation is observed in this case. Since the slope has a positive sign, the correlation might correspond to the strong adsorption region, but this is not the case since calculated Cu-O_2 binding energies increase with the redox potential. In contrast to metal phthalocyanines and metal porphyrins, $\theta_{Cu(I)}$ is practically constant and equal to 1 since kinetic measurements were conducted at potentials well below the Cu(II)/(I) formal potentials of the catalysts examined. This will be discussed in detail in section 2.1.2.3.

Figure 12 shows a correlation for a series of Mn porphyrins for ORR versus the Mn(III)/(II) redox potential of the complexes. Again, the correlation has the shape of a volcano. Since as a parameter of activity the half-wave potential was used, it is difficult to estimate if the falling region is attributed to a damping effect of $\theta_{Mn(II)}$. However, the values of $E_{1/2}$ are well above the redox potential of the complexes.

2.1.2.2 Pyrolysed catalysts for ORR

As pointed out above, intact molecular catalysts like the metal complexes described above are not stable in fuel cell electrolytes so a whole new area of research started in the 1970s. The aim was to obtain pyrolytic materials from heat-treated MN4 complexes or other ingredients involving metal salts and carbon- and nitrogen-containing compounds [18, 29–31, 36–38]. Heat treatments of up to 1000° C have been used, and the structure of the obtained catalysts is still a matter of

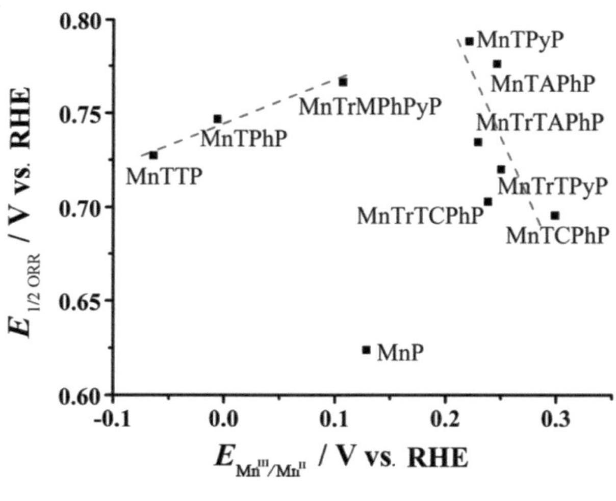

Figure 12.
Volcano correlation of half-wave potential $E_{1/2}$ ORR versus the $E_{Mn(III)/(II)}$ [19] (reproduced by permission of Wiley).

debate. These materials are not only more stable but importantly also more active. The effects of the heat treatment on the activity were interpreted by van Veen et al. in terms of the M(III)/(MII) redox potential [29, 30]. For example, when Fe porphyrins are pyrolysed, the ligand is destroyed by the pyrolysis, and the resulting surrounding ligand is much more electron-withdrawing in character. This results in a dramatic shift in the Fe(III)/(II) redox potential to more positive values. Literally, with heat treatment, the catalyst climbs the volcano correlation towards the top. Catalysts prepared without using a MN4 complex but other ingredients also exhibit more positive redox potentials than intact metal complexes. Many very highly active pyrolysed MNx catalysts do not exhibit visible redox signals in cyclic voltammograms, especially those prepared at higher temperatures, so the redox potential as predictor is lost. This could be attributed to the heterogeneity of active sites that can be formed at higher temperatures. Many authors have studied the changes occurring in carbon-supported metalloporphyrins and metallo-phthalocyanines. A variety of physical techniques have been used including X-ray photoelectron spectroscopy (XPS), Fourier-transform infrared spectroscopy, electron spin resonance and Mössbauer spectrometry and EXAFS [30].

When comparing the catalytic activity of a series of pyrolysed catalysts reported versus the M(III)/(II) redox potential, linear correlations are obtained which might correspond to incomplete volcano correlations (**Figure 13**) [39]. The slope is again +0.120 V/decade indicating that the equations derived above do explain both the experimental data of intact MN4 catalysts and pyrolysed catalysts showing once more that the redox potential is a rather universal reactivity descriptor for ORR and

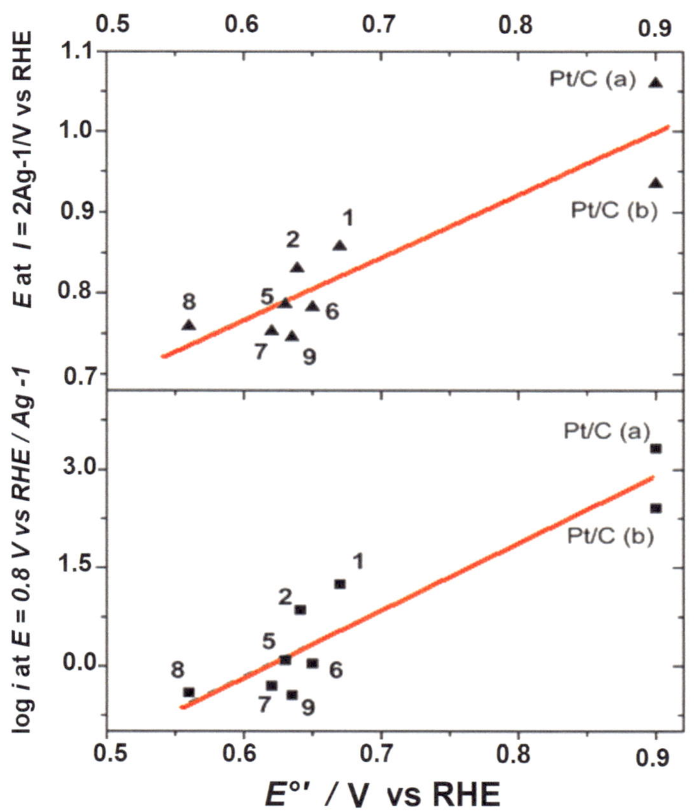

Figure 13.
Linear correlations of E at constant current and $(\log i)_E$ versus $E_{M(III)/(II)}$ for a series of pyrolysed MNx catalyst for ORR in acid media [39] (reproduced by permission of The Electrochemical Society).

possibly for other reactions [12, 13, 26]. The literature in this subject is very abundant, and more details and discussion about this pyrolysed MNx catalysts are beyond the scope of this chapter. However, it is important to remark that the redox potential is a reactivity predictor for this very important family of catalysts.

2.1.2.3 Cu non-macrocyclic complexes

In nature copper-containing laccase is an enzyme that under certain conditions can promote the four-electron reduction of O_2 to water without the formation of any peroxide. Laccase only works at a very narrow pH range, and its large molecular size prevents high current densities. The reaction can occur at very low overpotentials, and this is indeed very unusual in ORR catalysts. For this reason, some authors [40–43] have explored the catalytic activity for ORR of simpler Cu complexes. For copper complexes, the active state is Cu(I). We will focus our discussion on the effect of the redox potential. The reactivity trends of metal phthalocyanines and metal porphyrins illustrated in **Figures 5** and **6** show that CuN4 complexes exhibit very low activity for ORR. One of the reason for the low activity is that these Cu complexes are in the oxidation state Cu(II) and cannot be reduced to Cu(I) due to the rigidity of the planar phthalocyanine ligand since the reduction process involves a change in geometry around the Cu centre from planar Cu(II) to tetrahedral (Cu(I) (see **Figure 14**). The other reason is that CuPc has no frontier orbital with *d*-metal character that can bind O_2. This was illustrated in

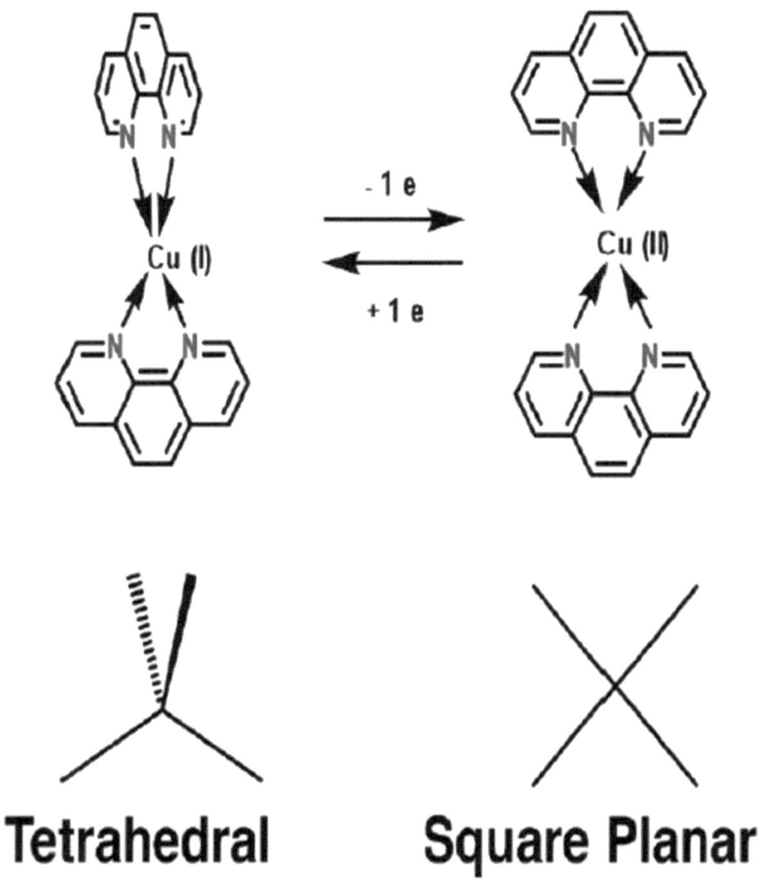

Figure 14.
Illustration of changes in geometry when Cu(I) is oxidized to Cu(II) in a $Cu(phen)_2$ complex.

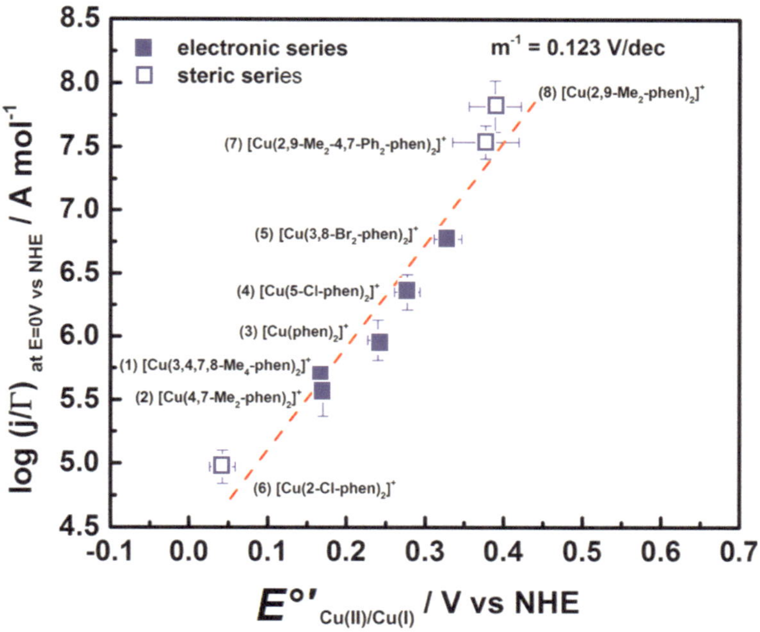

Figure 15.
Plot of log(i/Γ) at E = 0.0 V vs. NHE versus the Cu(II)/(I) redox potential of the complex for ORR [44] (reproduced by permission of the American Chemical Society).

Figure 6. Cu phenanthrolines are flexible, and then redox processes can occur on the metal centre.

The Cu(II)/(I) redox potential of copper phenanthroline is a reactivity descriptor for ORR. **Figure 15** shows a linear correlation between log i and $E°$ with a slope close to +0.120 V/decade which seems to be an incomplete volcano correlation [44].

3. Conclusions

The redox potential of the catalyst plays different roles in electrochemistry. Depending on whether the catalyst is present in the homogeneous phase or anchored or adsorbed on the electrode surface, the correlations between activity and redox potential can be different. For outer-sphere reactions occurring in the solution phase, the catalytic activity as log k versus the redox potential of the catalyst $E°$ increases linearly with the redox potential, whereas for inner-sphere chemical catalysts, the activities observed are higher than those predicted by the redox potential of the catalyst.

For reactions promoted by molecular catalysts immobilized on the electrode surface, correlations of $(\log i)_E$ versus $E°$ are observed. Linear and volcano correlations are obtained. When comparing MN4 macrocyclic complexes having a wide range of redox potentials, typical volcano correlations are generally observed in electrocatalysis. In classical volcano correlations, the different electrodes are metals or alloys. In this case the binding energy of key intermediates is used as reactivity descriptor. The activity gradually increases with the binding energy up to a point. Beyond that point the activity decreases with larger binding energies as the surface becomes covered with adsorbed intermediates. Similar correlations are observed with MN4 molecular catalysts. Further, with both metals and molecular catalysts, the strong adsorption region of the volcano involves the four-electron reduction of O_2, whereas the weak adsorption region includes the two-electron reduction catalysts. The decrease in activity for ORR using MN4 metal complexes does not seem to be related to a gradual occupation of the active site but rather to a gradual decrease in the amount of M(II) active sites. This is observed for those catalysts that have M(III)/(II) redox potentials more negative than the electrode potential chosen for comparing the activities. Cu phenanthroline complexes follow similar correlations. It is observed that the activity increases as the Cu(II)/(I) redox potential increases, showing only a linear correlation.

A general conclusion from electrocatalytic phenomena is that if volcano correlations are well established, then for a particular reaction the properties of the catalyst can be "tuned" so to improve their activity. The optimal properties can involve many other parameters such as metal-to-metal separation, crystal orientation, stability, alloying, nanostructure and redox potential of catalyst, so this is an open field for both experimentalists and theoreticians to find the ways of improving the catalytic activity of electrode surfaces.

The implications of future development in this area will have a tremendous impact in energy conversion devices, electrosynthesis and electrochemical sensors, just to mention a few.

Acknowledgements

This work has been financed by the Fondecyt Project 1181037, Fondecyt Project 1171408 and 1140199 and Núcleo Milenio de Ingeniería Molecular P07-006.

Author details

José H. Zagal*, Ingrid Ponce and Ruben Oñate
Department de Chemistry of Materials, Department of Chemistry of the Environment, Faculty of Chemistry and Biology, Laboratory of Electrocatalysis, University of Santiago de Chile, Santiago, Chile

*Address all correspondence to: jose.zagal@usach.cl

IntechOpen

© 2019 The Author(s). Licensee IntechOpen. This chapter is distributed under the terms of the Creative Commons Attribution License (http://creativecommons.org/licenses/by/3.0), which permits unrestricted use, distribution, and reproduction in any medium, provided the original work is properly cited.

References

[1] Appleby AJ, Zagal JH. Free energy relationships in electrochemistry: A history that started in 1935. Journal of Solid State Electrochemistry. 2011;**15**: 1811-1832. DOI: 10.1007/s10008-011-1394-8

[2] Frumkin A. Bemerkung zur Theorie der Wasserstoffüberspannung. Zeitschrift für Physikalische Chemie. 1932;**160A**:116-118. DOI: 10.1515/zpch-1932-16011

[3] Bockris JO, Khan SUM. Some experimental techniques. In: Surface Electrochemistry. Boston, MA: Springer US; 1993. pp. 1-58. DOI: 10.1007/978-1-4615-3040-4_1

[4] Tafel J. Über die Polarisation bei kathodischer Wasserstoffentwicklung. Zeitschrift für Physikalische Chemie. 1905;**50**:641-712. DOI: 10.1515/zpch-1905-5043

[5] Petrii OA, Nazmutdinov RR, Bronshtein MD, Tsirlina GA. Life of the Tafel equation: Current understanding and prospects for the second century. Electrochimica Acta. 2007;**52**: 3493-3504. DOI: 10.1016/j.electacta.2006.10.014

[6] Zagal JH, Koper MTM. Reactivity descriptors for the activity of molecular MN4 catalysts for the oxygen reduction reaction. Angew Chemie. 2016;**55**: 14510-14521. DOI: 10.1002/anie.201604311

[7] Savéant J-M. Elements of Molecular and Biomolecular Electrochemistry. Hoboken, NJ, USA: John Wiley & Sons, Inc.; 2006. DOI: 10.1002/0471758078

[8] Lexa D, Mispelter J, Savéant JM. Electroreductive alkylation of iron in porphyrin complexes. Electrochemical and spectral characteristics of Sigma.-alkylironporphyrins. Journal of the American Chemical Society. 1981;**103**: 6806-6812. DOI: 10.1021/ja00413a004

[9] Lexa D, Savéant JM, Wang DL. Electroreductive alkylation of iron porphyrins. Iron(III), iron(II) and iron (I) alkyl complexes from the reaction of doubly reduced iron(II) porphyrins with alkyl halides. Organometallics. 1986;**5**:1428-1434. DOI: 10.1021/om00138a022

[10] Lexa D, Savéant JM, Su KB, Wang DL. Chemical vs. redox catalysis of electrochemical reactions. Reduction of trans-1,2-dibromocyclohexane by electrogenerated aromatic anion radicals and low oxidation state metalloporphyrins. Journal of the American Chemical Society. 1987;**109**: 6464-6470. DOI: 10.1021/ja00255a036

[11] Lexa D, Savéant JM, Schaefer HJ, Khac Binh S, Vering B, Wang DL. Outer-sphere and inner-sphere processes in reductive elimination. Direct and indirect electrochemical reduction of vicinal dibromoalkanes. Journal of the American Chemical Society. 1990;**112**:6162-6177. DOI: 10.1021/ja00173a002

[12] Zagal JH. Metallophthalocyanines as catalysts in electrochemical reactions. Coordination Chemistry Reviews. 1992; **119**:89-136. DOI: 10.1016/0010-8545(92)80031-L

[13] Zagal JH, Griveau S, Silva JF, Nyokong T, Bedioui F. Metallophthalocyanine-based molecular materials as catalysts for electrochemical reactions. Coordination Chemistry Reviews. 2010;**254**: 2755-2791. DOI: 10.1016/j.ccr.2010.05.001

[14] Masa J, Ozoemena K, Schuhmann W, Zagal JH. Oxygen reduction reaction using N4 - metallomacrocyclic catalysts:

Fundamentals on rational catalyst design. Journal of Porphyrins and Phthalocyanines. 2012;**16**:761-784. DOI: 10.1142/S1088424612300091

[15] Boulatov R. Billion-year-old oxygen cathode that actually works: Respiratory oxygen reduction and its biomimetic analogs. In: Zagal JH, Bedioui F, Dodelet J-P, editors. N4-Macrocyclic Metal Complexes. New York, NY: Springer New York; 2006. pp. 1-40. DOI: 10.1007/978-0-387-28430-9_1

[16] Zagal JH, Páez MA, Silva JF. Fundamental aspects on the catalytic activity of metallomacrocyclics for the electrochemical reduction of O2. In: Zagal JH, Bedioui F, Dodelet J-P, editors. N4-Macrocyclic Metal Complexes. New York, NY: Springer New York; 2006. pp. 41-82. DOI: 10.1007/978-0-387-28430-9_2

[17] Jasinski R. A new fuel cell cathode catalyst. Nature. 1964;**201**:1212-1213. DOI: 10.1038/2011212a0

[18] Dodelet J-P. Oxygen reduction in PEM fuel cell conditions: Heat-treated non-precious metal-N4 macrocycles and beyond. In: Zagal JH, Bedioui F, Dodelet J-P, editors. N4-Macrocyclic Metal Complexes. New York, NY: Springer New York; 2006. pp. 83-147. DOI: 10.1007/978-0-387-28430-9_3

[19] Masa J, Schuhmann W. Systematic selection of metalloporphyrin-based catalysts for oxygen reduction by modulation of the donor-acceptor intermolecular hardness. Chemistry: A European Journal. 2013;**19**: 9644-9654. DOI: 10.1002/chem.201203846

[20] Hipps KW, Lu X, Wang XD, Mazur U. Metal d-orbital occupation-dependent images in the scanning tunneling microscopy of metal phthalocyanines. The Journal of Physical Chemistry. 1996;**100**: 11207-11210. DOI: 10.1021/jp960422o

[21] Wang Y, Wu K, Kröger J, Berndt R. Review article: Structures of phthalocyanine molecules on surfaces studied by STM. AIP Advances. 2012;**2**: 041402. DOI: 10.1063/1.4773458

[22] Cárdenas-Jirón GI, Gulppi MA, Caro CA, del Río R, Páez M, Zagal JH. Reactivity of electrodes modified with substituted metallophthalocyanines. Correlations with redox potentials, Hammett parameters and donor-acceptor intermolecular hardness. Electrochimica Acta. 2001;**46**:3227-3235. DOI: 10.1016/S0013-4686(01)00614-4

[23] van Veen JAR, van Baar JF, Kroese KJ. Effect of heat treatment on the performance of carbon-supported transition-metal chelates in the electrochemical reduction of oxygen. Journal of the Chemical Society, Faraday Transactions. 1981;**77**: 2827-2843. DOI: 10.1039/f19817702827

[24] Nørskov JK, Rossmeisl J, Logadottir A, Lindqvist L, Kitchin JR, Bligaard T, et al. Origin of the overpotential for oxygen reduction at a fuel-cell cathode. The Journal of Physical Chemistry. B. 2004;**108**: 17886-17892. DOI: 10.1021/jp047349j

[25] Zagal JH, Javier Recio F, Gutierrez CA, Zuñiga C, Páez MA, Caro CA. Towards a unified way of comparing the electrocatalytic activity MN4 macrocyclic metal catalysts for O2 reduction on the basis of the reversible potential of the reaction. Electrochemistry Communications. 2014;**41**:24-26. DOI: 10.1016/j.elecom.2014.01.009

[26] Silva N, Calderón S, Páez MA, Oyarzún MP, Koper MTM, Zagal JH. Probing the $Fe^{n+}/Fe^{(n-1)+}$ redox potential of Fe phthalocyanines and Fe porphyrins as a reactivity descriptor in the electrochemical oxidation of cysteamine. Journal of Electroanalytical Chemistry. 2018;**819**:502-510. DOI: 10.1016/J.JELECHEM.2017.12.068

[27] Zagal JH, Aguirre MJ, Páez MA. O_2 reduction kinetics on a graphite electrode modified with adsorbed vitamin B12. Journal of Electroanalytical Chemistry. 1997;**437**:45-52. DOI: 10.1016/S0022-0728(97)00253-2

[28] Cárdenas-Jirón GI, Zagal JH. Donor–acceptor intermolecular hardness on charge transfer reactions of substituted cobalt phthalocyanines. Journal of Electroanalytical Chemistry. 2001;**497**:55-60. DOI: 10.1016/S0022-0728(00)00434-4

[29] van Veen JAR, Colijn HA, van Baar JF. On the effect of a heat treatment on the structure of carbon-supported metalloporphyrins and phthalocyanines. Electrochimica Acta. 1988;**33**:801-804. DOI: 10.1016/S0013-4686(98)80010-8

[30] Bouwkamp-Wijnoltz AL, Visscher W, van Veen JAR, Boellaard E, van der Kraan AM, Tang SC. On active-site heterogeneity in pyrolyzed carbon-supported iron porphyrin catalysts for the electrochemical reduction of oxygen: An in situ Mössbauer study. The Journal of Physical Chemistry. B. 2002; **106**:12993-13001. DOI: 10.1021/jp0266087

[31] Chen Z, Higgins D, Yu A, Zhang L, Zhang J. A review on non-precious metal electrocatalysts for PEM fuel cells. Energy & Environmental Science. 2011;**4**:3167. DOI: 10.1039/c0ee00558d

[32] Randin J-P. Interpretation of the relative electrochemical activity of various metal phthalocyanines for the oxygen reduction reaction. Electrochimica Acta. 1974;**19**:83-85. DOI: 10.1016/0013-4686(74)85060-7

[33] Beck F. The redox mechanism of the chelate-catalyzed oxygen cathode. Journal of Applied Electrochemistry. 1977;**7**:239-245. DOI: 10.1007/BF00618991

[34] Ulstrup J. Catalysis of the electrochemical reduction of molecular oxygen by metal phthalocyanines. Journal of Electroanalytical Chemistry. 1977;**79**:191-197. DOI: 10.1016/S0022-0728(77)80411-7

[35] Zagal JH, Gulppi M, Isaacs M, Cárdenas-Jirón G, Aguirre MJ. Linear versus volcano correlations between electrocatalytic activity and redox and electronic properties of metallophthalocyanines. Electrochimica Acta. 1998;**44**:1349-1357. DOI: 10.1016/S0013-4686(98)00257-6

[36] Zhang L, Zhang J, Wilkinson DP, Wang H. Progress in preparation of non-noble electrocatalysts for PEM fuel cell reactions. Journal of Power Sources. 2006;**156**:171-182. DOI: 10.1016/J.JPOWSOUR.2005.05.069

[37] Bezerra CWB, Zhang L, Liu H, Lee K, Marques ALB, Marques EP, et al. A review of heat-treatment effects on activity and stability of PEM fuel cell catalysts for oxygen reduction reaction. Journal of Power Sources. 2007;**173**:891-908. DOI: 10.1016/J.JPOWSOUR.2007.08.028

[38] Song C, Zhang J. Electrocatalytic oxygen reduction reaction. In: Zhang J, editor. PEM Fuel Cell Electrocatalysts and Catalyst Layers. London: Springer London; 2008. pp. 89-134. DOI: 10.1007/978-1-84800-936-3_2

[39] Zagal JH, Ponce I, Baez D, Venegas R, Pavez J, Paez M, et al. A possible interpretation for the high catalytic activity of heat-treated non-precious metal Nx/C catalysts for O2 reduction in terms of their formal potentials. Electrochemical and Solid-State Letters. 2012;**15**:B90. DOI: 10.1149/2.032206esl

[40] Thorseth MA, Tornow CE, Tse ECM, Gewirth AA. Cu complexes that catalyze the oxygen reduction reaction. Coordination Chemistry

Reviews. 2013;**257**:130-139. DOI: 10.1016/j.ccr.2012.03.033

[41] McCrory CCL, Ottenwaelder X, Stack TDP, Chidsey CED. Kinetic and mechanistic studies of the electrocatalytic reduction of O_2 to H_2O with mononuclear Cu complexes of substituted 1,10-phenanthrolines. The Journal of Physical Chemistry. A. 2007; **111**:12641-12650. DOI: 10.1021/jp076106z

[42] Solomon EI, Augustine AJ, Yoon J. O_2 reduction to H_2O by the multicopper oxidases. Dalton Transactions. 2008;**30**: 3921. DOI: 10.1039/b800799c

[43] Lei Y, Anson FC. Dynamics of the coordination equilibria in solutions containing copper(II), copper(I), and 2,9-dimethyl-1,10-phenanthroline and their effect on the reduction of O_2 by Cu (I). Inorganic Chemistry. 1995;**34**: 1083-1089. DOI: 10.1021/ic00109a014

[44] Venegas R, Muñoz-Becerra K, Lemus LA, Toro-Labbé A, Zagal JH, Recio FJ. Theoretical and experimental reactivity predictors for the electrocatalytic activity of copper phenanthroline derivatives for the reduction of dioxygen. Journal of Physical Chemistry C. 2019;**332**: 19468-19478. DOI: 10.1021/acs.jpcc.9b03200

Chapter 5

Effects of Electrolyte Additives on Nonaqueous Redox Flow Batteries

Qian Xu, Chunzhen Yang and Huaneng Su

Abstract

The widespread utilization of nonaqueous redox flow batteries is hindered by the low performance. Including some kinds of additives in electrolyte is a possible and facile solution. In this chapter, the effects of carbon dioxide gas, EC/DMC, and antimony ions on the electrochemical performance of nonaqueous redox flow batteries are disclosed. The results show that the ohmic resistance of the deep eutectic solvent (DES) electrolyte reduces significantly when adding carbon dioxide gas and EC/DMC, the percentage of reduction increases with the volume percentage of EC/DMC in electrolyte, and the reaction kinetics almost keeps unchanged for carbon dioxide gas and EC/DMC additives. For the additive of antimony ions, the electrochemical reaction kinetics of active redox couple is enhanced, the diffusion coefficient of active ions also increases, and the charge transfer resistance decreases. The antimony ions electrodeposited on the surface of graphite felt contribute a catalytic effect on the electrochemical reaction so as to improve the performance. However, due to the trade-off between the enhanced kinetics and reduced active surface area, the optimum concentration of antimony ions is found to be 15 mM. In addition, the flow battery assembled with negative electrolyte containing antimony ions exhibits 31.2% higher power density than that of pristine DES electrolyte.

Keywords: redox flow batteries, deep eutectic solvent (DES), electrolyte additives, carbon dioxide, EC/DMC, antimony ions

1. Introduction

Recent years, with the development of energy storage technology, people prefer to use the redox flow battery (RFB) in large-scale energy storage. RFB has outstanding advantages: it has no limitation by the geographical environment and the sites, the design of cell structure is flexible, and also it shows rapid response to charge and discharge switching and with a long cycle life [1–3]. The aqueous electrolytes for RFB are mostly used during the past decades. However, such systems have a very narrow operating potential window (<2 V) due to the effects of water decomposition which limit the potential power output [4]. In recent years, the research of flow battery technology has extended from aqueous system to nonaqueous system in order to obtain higher potential window. Organic solvents have a much higher electrochemical window, e.g., 5.0 V for acetonitrile (CH_3CN), so that people can gain much higher energy output and power [5]. Even so, organic solvents show potential safety hazards because of their volatility, toxicity, and flammability; moreover, moisture or oxygen contamination can also adversely affect battery performance [6]. To this point, some ionic liquids (ILs) have advantages to solve the problem.

IntechOpen

Ionic liquids are salts which can melt at room temperature. These salts have a potential window as wide as organic solvents. Compared with other solvents, ionic liquids have high thermal and electrochemical stability, and the conductivity is higher than aqueous electrolytes [7]. Because of these advantages, ILs have been applied to a lot of fields, such as lithium-ion batteries [8], dye-sensitized solar cells [9], electrolytes in sensors [10], electrochemical capacitors [11], lead acid batteries [12], and fuel cells [13], and even applied to flow batteries recently as electrolyte solutions [14]. In 2015, the applicability of ionic liquids has been explored by Ejigu et al. as solvents of metal complex-containing redox flow battery [15]. Zhang et al. applied $TEAPF_6$ and $EMIPF_6$, two ionic liquids, to nonaqueous redox flow batteries. The results of charge and discharge tests showed that the coulombic efficiencies range from 43.46 to 57.44% [16]. However, there still exist some problems which have become the limitation for their large-scale applications, such as complex synthesis steps, cost, and availability.

The deep eutectic solvent (DES) can be recognized as a peculiar ionic liquid; it is consisting of a conjunction with a stoichiometric ratio of acceptors of hydrogen bond (like quaternary ammonium salts) and donors of hydrogen bond (like compounds of amides, carboxylic acids, and polyols) into the eutectic mixture. We can prepare and use DES under ambient conditions; low cost is the main advantage of DES, which is cheaper than conventional ionic liquids to an order of magnitude; DES is easy to prepare and its overall biodegradability is an advantage too [17–20]. People have carried out preliminary works on the DES. Lloyd et al. had studied the kinetics of electron transfer of the Cu(I)/Cu(II) redox couple with chronoamperometry at a platinum electrode; they named cyclic voltammetry and impedance spectroscopy in a deep eutectic solvent, which consist of choline chloride and ethylene glycol as ethaline [21]. Thereafter, an all-copper hybrid redox flow battery was demonstrated in the ethaline DES [22]. Nevertheless, because of the mass transport limitations and the poor electrolyte conductivity, the energy efficiencies could only reach to 52 and 62% when the current densities were 10 and 7.5 mA cm^{-2}, respectively. In recent researches, the electrochemical and transport characteristics of Fe(II)/Fe(III) as well as V(II)/V(III) redox couples in the DES electrolytes had been studied by Xu group [23, 24].

When people use pristine DES as electrolyte, the main issues are large viscosity and its small diffusion coefficient, which will lead to large pumping loss and cause low efficiency for RFBs [25]. In order to overcome these problems, researchers proposed ways of adding appropriate additives into the electrolyte, like gas and ionic additives, and tuning the active materials by adopting molecular engineering [26, 27]. In 2015, the viscosity changes of ionic liquids after adding SO_2 had been studied by Zeng et al. [27]. The results indicate that when the concentration of SO_2 increases, the viscosity of conventional ionic liquids will decrease sharply. Also, some studies have shown that the adoption of high-pressure CO_2 in DES can certainly improve the physical properties of the RFB system, and this can also increase its electrical conductivity [28–32]. However, there are not too much literature about how the gas additives influence on the electrochemical performance of nonaqueous flow batteries. Moreover, some supporting electrolytes are widely used in lithium-ion batteries for the reduction of electrolyte viscosity as well as the enhancement of cell performance [33], but few are reported in redox flow batteries.

Metal ions are also widely used as additives of electrolytes for electrochemical energy system. Antimony (Sb) has the advantages of low cost, good chemical stability, and high catalytic activity, such that it is widely used in the field of electrocatalysis and battery [23, 34]. In 2015, Shen et al. introduced $SbCl_3$ into a vanadium redox flow battery (VRFB) [35]. The work shows that the added $SbCl_3$ can improve the electrochemical activity and redox kinetics of V(III)/V(II).

In this chapter, the effects of electrolyte additives on nonaqueous redox flow batteries are introduced. The additives include CO_2 gas, ethyl acetate/dimethyl carbonate (EC/DMC) supporting electrolyte, and antimony ions. The effects were studied by means of viscosity test, cyclic voltammetry (CV), electrochemical impedance spectroscopy (EIS), and charge-discharge tests. The results here disclose an effective and convenient approach to improve the cell performance of nonaqueous redox flow batteries.

2. Materials, components, and operation parameters

2.1 Preparation of electrolyte

The DES electrolyte could be prepared by combining choline chloride (Aladdin, 98%) and urea (Sinopharm Reagent 57-13-6, 99%), and the molar ratio is 1:2. The solution at 120°C with a magnetic stirrer is heated and stirred until it formed a colorless and transparent liquid, this mixture is also known as "reline," and the temperature of liquid was also reduced to the room temperature. With long-term placement, the DES would appear some white crystalline precipitate, so it needed to be heated over 50°C in advance of the experiment; this process should proceed on the magnetic stirrer for near 30 minutes, and the rotor's rotation promotes the crystalline precipitate's dissolution. When finishing the experiment, DES should be kept in the sealed glass jar timely to avoid pollution caused by water vapor and oxygen in the air. The active material $FeCl_3$ (Sinopharm Reagent 7705-08-0, 99%) with a concentration of 0.1 mol L^{-1} was added into reline, and the mixture at 120°C was heated and stirred to obtain the electrolyte. Vacuum dried the prepared electrolyte for 24 hours before the start of experiment. Divide the solution into two portions, one portion serving as the pristine and the other one being fed with 0.1 MPa of CO_2 (with the purity of 99.99%). The EC (Macklin 98%) and DMC (Macklin 99%) mixture were prepared by 1:1 vol.% and stirred evenly when adding into the DES electrolyte. For the ion additive, the $SbCl_3$ (99%, Sinopharm Chemical Reagent Co., Ltd.) was added into the negative DES electrolyte at a concentration of 5, 10, 15, and 20 mM, respectively.

2.2 Viscosity measurement

Viscosity measurements were conducted with a "DV-2 + PRO" digital viscometer (Shanghai Nirun Co., Ltd.). The electrolyte was placed on a thermostatic dry heater, and the temperature of the solution was measured by a thermocouple thermometer. When the electrolyte reached the predetermined temperature, measure three times the viscosity at each temperature point, and take the average.

2.3 Electrochemical characterizations

Cyclic voltammetry (CV) and electrochemical impedance spectroscopy (EIS) were operated on Chenhua® CHI600 electrochemical workstation. These measurements used a traditional three-electrode system, its working electrode was a 5-mm-diameter glassy carbon electrode, a platinum electrode was used as counter electrode, and the reference electrode was a saturated calomel electrode together with a salt bridge which was filled with the saturated potassium chloride solution. Before each test, polish the glassy carbon electrode on the deerskin with the 0.2-mm aluminum powder, then place the polished electrode in deionized water, and clean it with ultrasonic waves. The CV scan was performed in the range of −0.7–0.9 V for the electrolyte in the case of with and without CO_2. Before the measurement, purge the electrolyte with nitrogen for 15 minutes in order to remove oxygen which dissolved in the electrolyte. The CVs were measured at 25, 35, and 45°C, respectively, and at each temperature the CV was tested three times. Simultaneously, in the electrochemical impedance spectroscopy measurements, the sinusoidal excitation voltage suitable for the cells was 5 mV. The frequency was in the range of 0.01 Hz–100 kHz. The potential was settled at 0.15 V in order to ensure similar polarization.

3. Effect of carbon dioxide additive

3.1 Cyclic voltammetry

The cyclic voltammograms of 0.1 M $FeCl_3$ in reline DES at different temperatures with and without 0.1 MPa CO_2 are shown in **Figure 1**. The result indicates that in the range of −0.7–0.9 V with the scan rate of 25 mV s^{-1}, there are two peaks: one is the oxidation peak and the other one is the reduction peak, and the position does not change basically where the peaks appear. For example, in the case of 25°C, the pristine electrolyte's oxidation and reduction peaks appeared at 0.32 and 0.032 V, respectively. After the CO_2 is being introduced, the electrolyte's oxidation peak appeared at 0.318 V and the reduction peak appeared at 0.033 V. This means that the introduction of CO_2 does not affect the electrolyte composition because there is no generation of new material and it does not substantially affect the redox reversibility of the electrolyte. (The physical absorption of CO_2 is also confirmed in reline DES.) On the basis of the redox peak currents, after the CO_2 has been introduced, the peak current does not change significantly. In the case of 25°C, the pristine electrolyte showed the oxidation peak current density of 0.176 mA cm^{-2}, and in this case, the reduction peak

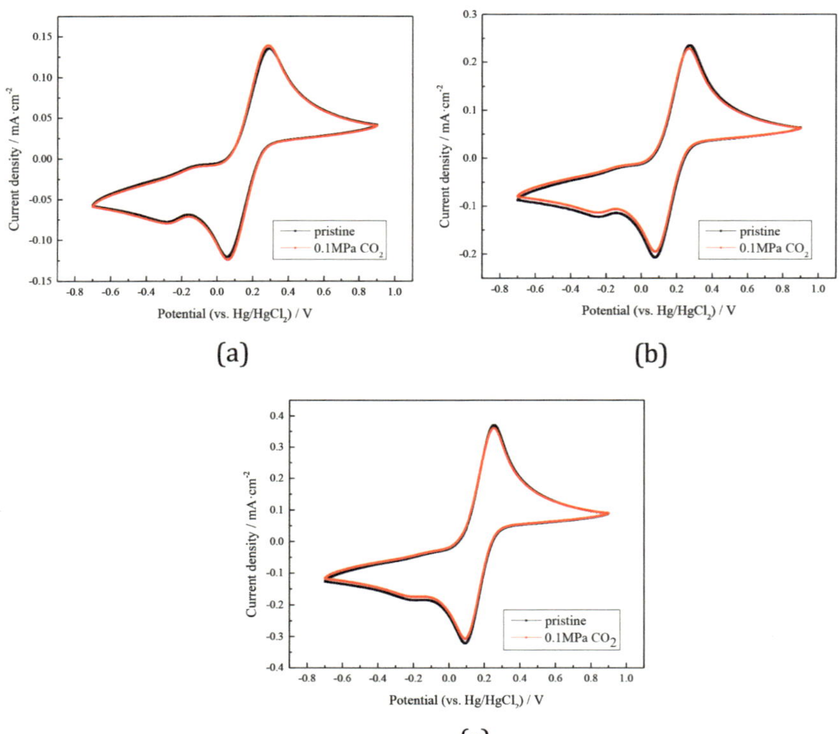

Figure 1.
Cyclic voltammograms of 0.1 M $FeCl_3$ in reline DES with and without 0.1 MPa CO_2 at different temperatures: (a) 25°C, (b) 35°C, and (c) 45°C.

current density was −0.158 mA cm^{-2}. After introducing CO_2, they became 0.178 and −0.16 mA cm^{-2}, respectively. The result shows that the rate of redox reaction of Fe(II)/Fe(III) redox couple in DES changes little when CO_2 is being added in.

3.2 Electrochemical impedance spectroscopy

Electrochemical impedance spectroscopy (EIS) was carried out to further investigate how the addition of CO_2 influences the electrochemical performance of Fe(III) ion in DES electrolyte. The Nyquist plots of Fe(III) in DES electrolyte at different temperatures with and without CO_2 are shown in **Figure 2**. Each plot shows a similar illustration: in the high-frequency region, there is a semicircle and in the low-frequency region, there is a straight line upward. They correspond to the transfer reaction of charge at the interface of electrode/electrolyte and the diffusion of iron species in the electrolyte, respectively, and this result suggests that electrochemical reaction and diffusion steps mix-control the Fe(III)/Fe(II) redox reaction [36]. There is a distance from the crossing of the semicircle's left end and the abscissa to the origin, which is the ohmic resistance of the electrolyte, and the semicircle's diameter is the electrochemical reaction resistance of the electrolyte. In order to determine the ohmic resistance and the electrochemical reaction resistance precisely before and after adding 0.1 MPa CO_2 to the reline DES which contains 0.1 M $FeCl_3$ in case of different temperatures, the data were fitted and the straight line in **Figure 2** showed the results. **Figure 2** shows a simplified equivalent circuit; the resistance of the ion migration process in the solution is represented as *Rs*, that is to say, the ohmic resistance of the solution. *Rt* is the resistance of electrochemical reaction, the resistance of the electron transfer step. *CPE* represents

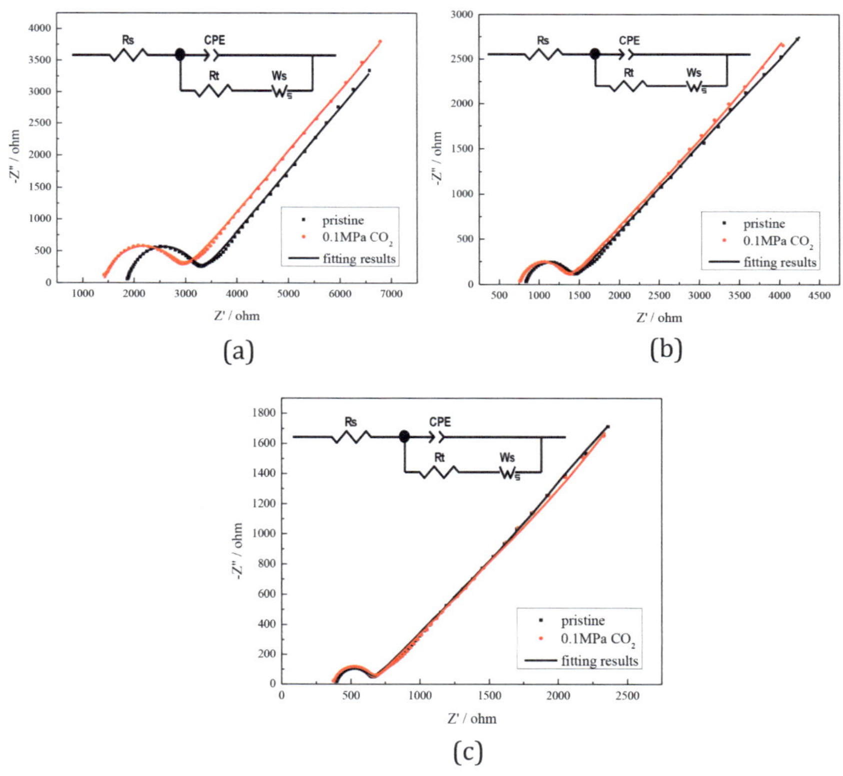

Figure 2.
Nyquist plots of the electrolyte with and without CO_2 at different temperatures and the corresponding equivalent circuit: (a) 25°C, (b) 35°C, and (c) 45°C.

Temperature (°C)	*Rs* (ohm)		*Rt* (ohm)	
	Pristine	0.1 MPa CO_2	Pristine	0.1 MPa CO_2
25	1867	1409	1337	1446
35	834.7	757.5	560.6	587.3
45	390.2	373.2	251.3	277.8
55	233.6	227.6	129.3	149

Table 1.
The parameters obtained from fitting the EIS plots with the equivalent circuit for carbon dioxide additive.

the double-layer capacitance of the interface, simulating the process of the double layer in case of charge and discharge. *Ws* stands for the concentration polarization impedance, which simulates the liquid phase's mass transfer.

The Z-view simulation helps to obtain the parameters of the equivalent circuit, which are listed in **Table 1**. It can be found that the resistance *Rs* of the electrolyte decreases after the addition of 0.1 MPa CO_2. In detail, at the temperature of 25°C, it decreases from 1867 to 1409 ohm. The solubility of CO_2 in the electrolyte will decrease when the temperature increases, so the decline percentage of *Rs* becomes small, and it is consistent with the trend of viscosity. What is more, the electrochemical reaction resistance of the solution slightly increases, implying that the addition of CO_2 would slow down the charge transfer process of the electrolyte solution. The redox flow battery with DES electrolyte adding CO_2 has a little-changed overall performance compared with the pristine DES electrolyte.

4. Effect of EC/DMC supporting electrolyte

4.1 Cyclic voltammetry

The cyclic voltammograms of 0.1 M $FeCl_3$ in reline DES without and with EC/DMC are shown in **Figure 3**. The volume of the tested DES electrolyte is 40 ml. The result indicates that with the scan rate of 50 mV s^{-1} in the range of −0.7–1.2 V, there appear only one oxidation peak and only one reduction peak. The position of the peak does not distinctly change with the addition of EC/DMC. After the introduction of EC/DMC (4 ml, 10% vol.), the oxidation peak and the reduction peak shift left and right for <50 mV, respectively. This suggests that the introduction of EC/DMC neither affects the electrolyte composition nor substantially affects the redox reversibility of the electrolyte. In terms of the redox peak currents, after the introduction of EC/DMC, the peak current increases for <10%. The result implies that the rate of redox reaction of Fe(II)/Fe(III) redox couple in DES does not change remarkably with the addition of EC/DMC.

4.2 Electrochemical impedance spectroscopy

The Nyquist plots of Fe(III) in DES electrolyte with and without EC/DMC are shown in **Figure 4**. For the pristine DES, the plot shows a semicircle in the high-frequency region, and in the low-frequency region, there is a straight line upward; the semicircle corresponds to the charge transfer reaction at the electrode/electrolyte interface, and the straight line upward corresponds to the diffusion of iron species in the electrolyte, suggesting electrochemical reaction and diffusion steps mix-control the Fe(III)/Fe(II) redox reaction. With the addition of EC/DMC,

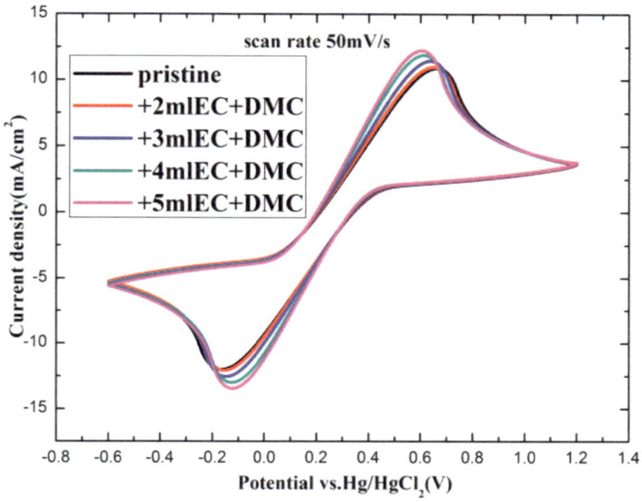

Figure 3.
CV curves of different concentrations of EC/DMC additive in DES with 0.1 mol L^{-1} *$FeCl_3$ with a scan rate of 50* $mV\,s^{-1}$.

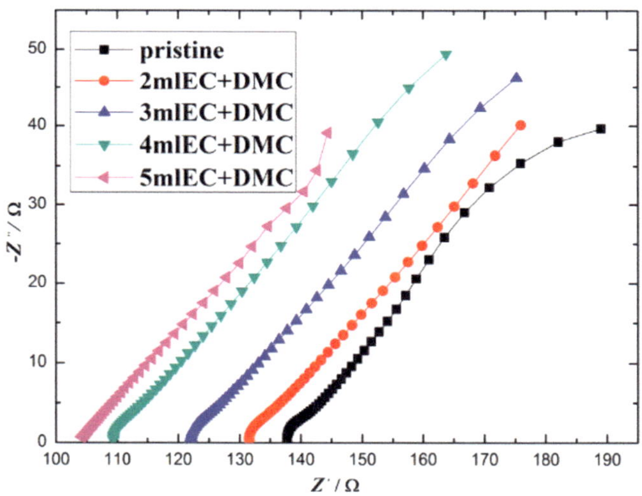

Figure 4.
AC impedance spectra for different concentrations of EC/DMC in DES with 0.1 mol L^{-1} *$FeCl_3$.*

the radius of semicircle reduces, suggesting the charge transfer reaction becomes diffusion-controlled. With addition of more EC/DMC, the ohmic resistance (*Rs*) of the DES electrolyte becomes smaller. For example, for the pristine DES, the *Rs* is 137.9 Ω; for the electrolyte with 3 ml EC/DMC, it is 122.2 Ω; and for the DES with 5 ml EC/DMC, it reduces to 103.4 Ω. The electrochemical reaction resistance (*Rt*) almost keeps unchanged with the addition of EC/DMC.

The Raman spectroscopy shows that the introduction of EC/DMC does not change the shape of spectrum significantly (**Figure 5**). The peak near 3000 cm^{-1} is a result of the overlap of the characteristic peaks of choline chloride and ethylene glycol (the main components of DES), while a new characteristic peak appears around 895 cm^{-1} when EC/DMC is added into the DES electrolyte, which can be attributed to the symmetrical stretching vibration of C-O-C bond when aliphatic ethers exist in the nonaqueous solution [33]. The reduction of ohmic resistance after the addition of EC/DMC should be the result of the stretching vibration of C-O-C bond.

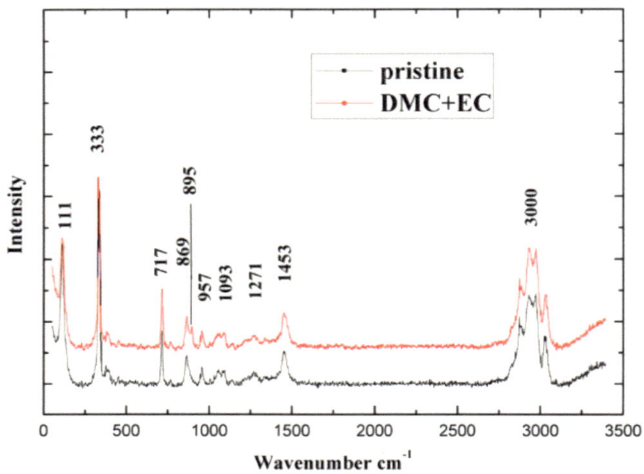

Figure 5.
Raman spectra of solvents without and with EC/DMC additive.

5. Effect of antimony ions

Inspired by the use of antimony ions as the additive in VRFBs, the same ions are tried to be the additive in DES electrolyte nonaqueous RFBs. The active materials used in negative side electrolyte are V(III)/V(II) redox ions.

5.1 Cyclic voltammetry

The cyclic voltammetry curves of negative electrolyte containing 0.1 mol L^{-1} VCl_3 with different concentrations of $SbCl_3$ are shown in **Figure 6**, and the scanning rate is 25 mV s^{-1}. It shows two obvious peaks corresponding to V(III)/V(II) redox couple. A small peak appears at the position of −0.2 V, and since there is no peak in DES, thus the small peak at −0.2 V should be caused by impurities in the raw materials. In addition, there is no new peak, which further proves that after

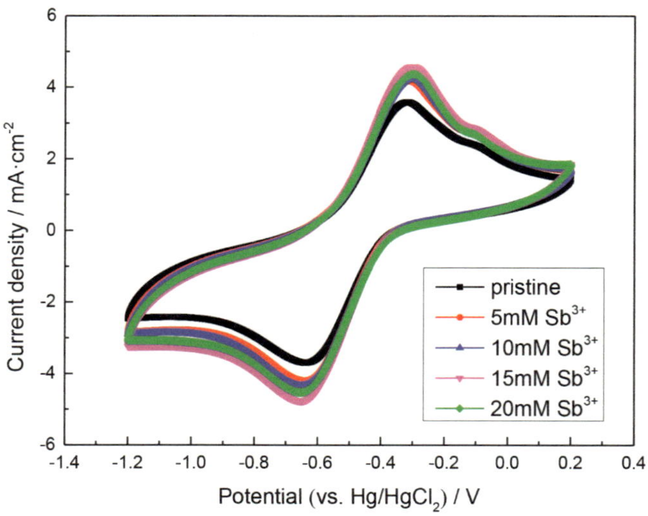

Figure 6.
Effect of Sb^{3+} on CV curves of 0.1 M VCl_3 in DES at the scanning rate of 25 mV s^{-1}.

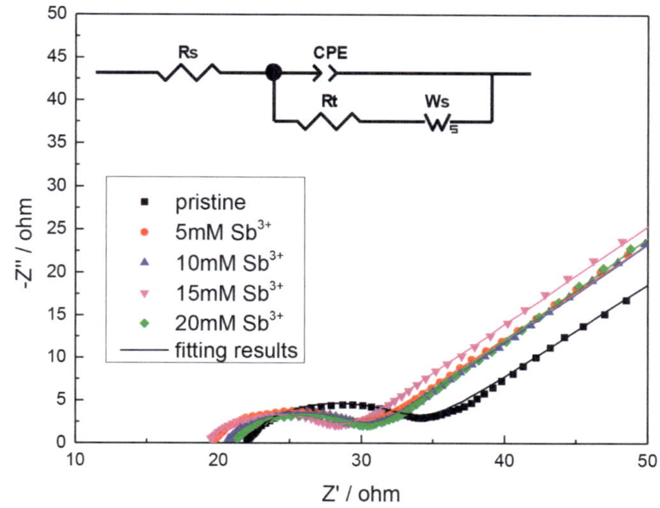

Figure 7.
Nyquist plot of 0.1 M VCl_3 in DES with different concentrations of Sb^{3+} ions.

the addition of Sb^{3+} ions, there is no chemical reaction with V(III) ions to generate a new substance. For the pristine electrolyte, the peak current densities were 3.583 and −3.691 mA cm^{-2}, respectively. After the introduction of Sb^{3+} ions, both of them increase and reach the maximum (4.589 and −4.764 mA cm^{-2}, respectively) when the concentration of Sb^{3+} ions is 15 mM. This indicates that the introduction of Sb^{3+} ions can accelerate the redox reaction rate of the battery and increase the collision between ions which makes it easier to overcome the activation energy and realize the electrochemical reaction [37].

5.2 Electrochemical impedance spectroscopy

The influence of $SbCl_3$ on the electrochemical properties of negative electrolyte was further investigated by EIS. The Nyquist plots of electrolyte without additive and with different concentration of $SbCl_3$ are shown in **Figure 7**. The combination of semicircle and the straight line upward suggests that the redox reaction of vanadium is mix-controlled by electrochemical polarization and concentration polarization [38]. **Figure 5** shows the corresponding simplified equivalent circuit, where *Ws* and *CPE* represent the concentration polarization impedance and double-layer capacitance of the solution, respectively.

It can be seen that the *Rs* and *Rt* of electrolyte decrease with the addition of different concentrations of $SbCl_3$ and reach the minimum when the concentration is 15 mM. The corresponding parameters of the equivalent circuit obtained by the Z-view simulation are shown in **Table 2**. The *Rs* and *Rt* of the solution with 15 mM $SbCl_3$ were the smallest; they were 19.41 and 8.95 ohm cm^{-2}, respectively, lower than that of the pristine electrolyte (22.03 and 11.57 ohm m^{-2}). The reduced electrochemical reaction resistance indicates that the charge transfer process of the electrolyte is accelerated, which reflects the higher electrochemical reaction rate. This is probably owing to the adhesion of Sb to the electrode and its catalytic effect. The increased *CPE* and *Ws* suggest that the Sb^{3+} in the electrolyte is able to promote the absorption and diffusion of V ions. The results further confirm that the addition of Sb^{3+} can improve the electrochemical reaction of V(III)/V(II).

In order to investigate the influence of the addition of $SbCl_3$ on the power density of the battery, the polarization curve of the flow battery was measured (the active material is $FeCl_3$ in positive side electrolyte). As shown in **Figure 8**, when

Concentration of Sb^{3+}	*Rs* (ohm cm^{-2})	*Rt* (ohm cm^{-2})	*CPE* (F cm^{-2})	*Ws*
Pristine	22.03	11.57	7.218×10^{-4}	0.5338
5 mM	20.72	9.89	9.86×10^{-4}	0.5381
10 mM	20.67	9.50	1.01×10^{-3}	0.5478
15 mM	19.41	8.95	1.15×10^{-3}	0.5581
20 mM	21.15	9.11	1.06×10^{-3}	0.5492

Table 2.
The parameters obtained from fitting the EIS plots with the equivalent circuit for antimony ion additive.

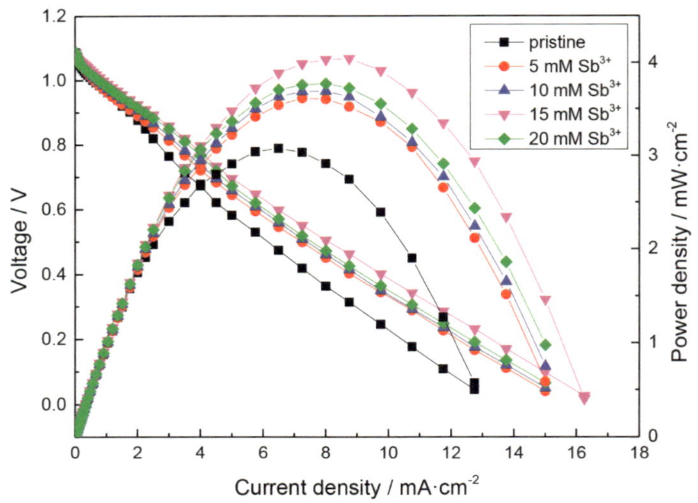

Figure 8.
Polarization curves of batteries with different concentrations of Sb^{3+} ions.

the concentration of Sb^{3+} ions increases, the maximum current density and maximum power density increase first and then decrease, and when the concentration reached 15 mM, they reached the maximum (16.25 mA cm^{-2} and 4.04 mW cm^{-2}, respectively), higher than those of pristine electrolyte (12.75 mA cm^{-2} and 3.08 mW cm^{-2}). From these experimental results, it can be seen that the electrochemical performance of V(III)/V(II) redox couple has been improved due to the adhesion of Sb ion to the electrode and its catalytic effect.

5.3 Physicochemical measurements

To explore the mechanism of Sb ions to improve the electrochemical performance of electrolytes, the surface morphology of graphite felts obtained after charging-discharging cycle was characterized by SEM. **Figure 9** shows that some particles adhere to the surface of graphite felt after the addition of $SbCl_3$.

With the increased concentration of $SbCl_3$, the particles on the surface of graphite felt increased. In order to analyze the specific composition of the observed ions, the energy-dispersive X-ray spectroscopy (EDX) was used to identify them. As shown in **Figure 10**, the transverse coordinate is the energy, and the vertical coordinate is the relative content of elements. The results of EDX show that the elements on the pristine surface of graphite felt are only C, O, V, and Cl and the particle detected after the addition of additives is Sb. Corresponding to the energy of 4.0 keV, the content of Sb on graphite felt increases gradually with the increased concentration of

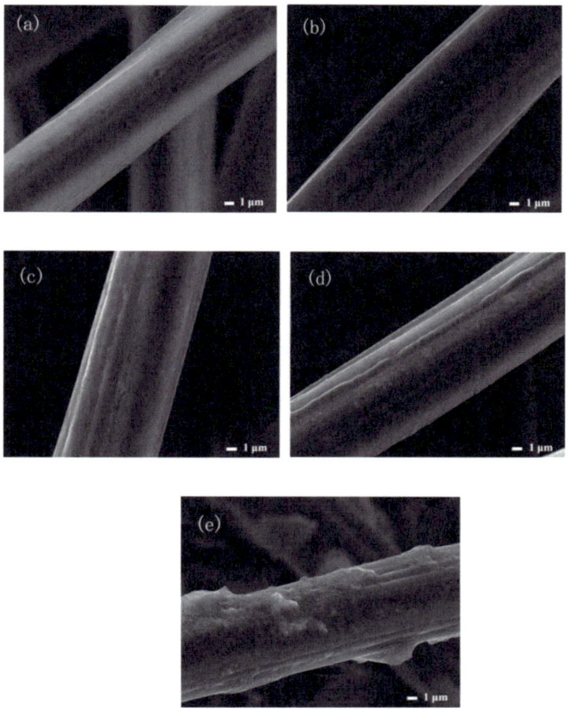

Figure 9.
The FESEM images of graphite felt electrode after cycling: (a) pristine, (b) 5 mM Sb^{3+}, (c) 10 mM Sb^{3+}, (d) 15 mM Sb^{3+}, and (e) 20 mM Sb^{3+}.

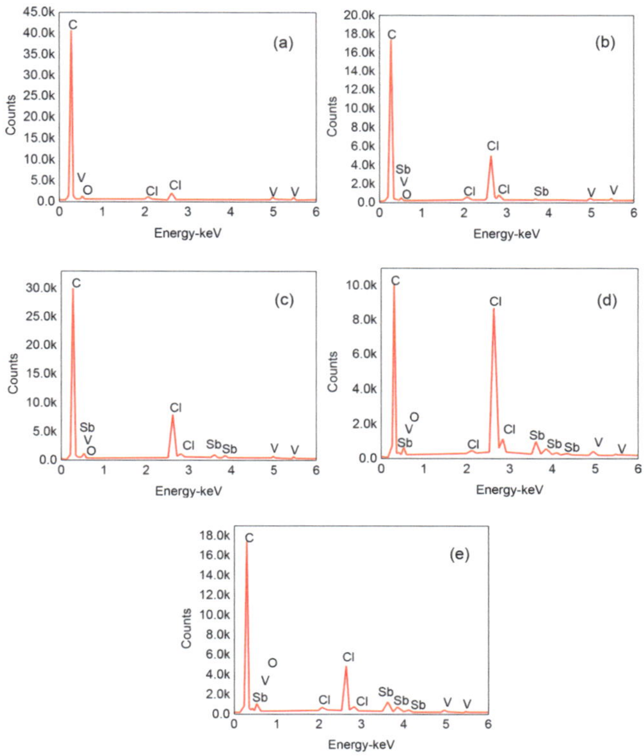

Figure 10.
EDX spectrogram of ions on the surface of graphite felts: (a) pristine, (b) 5 mM Sb^{3+}, (c) 10 mM Sb^{3+}, (d) 15 mM Sb^{3+}, and (e) 20 mM Sb^{3+}.

$SbCl_3$. The results suggest that the enhancement of the battery performance is owing to the catalytic effect of Sb ions. However, when the concentration was 20 mM, the accumulation of ions is more serious, which would result in partial pore blockage of graphite felt. Therefore, when the concentration of Sb^{3+} ions further increases, the electrochemical performance of the battery decreases slightly.

6. Conclusions

The effects of three kinds of additives including carbon dioxide gas, EC/DMC, and Sb^{3+} ions on the electrochemical performance of nonaqueous DES electrolyte redox flow batteries are explored. The ohmic resistance of the deep eutectic solvent (DES) electrolyte decreases significantly when adding carbon dioxide gas and EC/DMC, and the percentage of reduction increases with the volume percentage of EC/DMC in electrolyte, while for these two additives, the reaction kinetics almost keeps unchanged.

With CO_2 in DES, the electrochemical reaction resistance increases about 10%. For EC/DMC additive, the electrochemical reaction resistance almost keeps the same no matter the amount of additive in electrolyte. For the additive of Sb^{3+} ions in DES electrolyte, the electrochemical reaction kinetics of active redox couple is enhanced, the diffusion coefficient of active ions increases, and the charge transfer resistance decreases. The electrodeposited Sb^{3+} ions on electrode surface contribute a catalytic effect on the electrochemical reaction. However, due to the trade-off between the enhanced kinetics and reduced active surface area, the optimum concentration of Sb^{3+} ions is found to be 15 mM. In addition, the flow battery assembled with negative electrolyte containing Sb^{3+} ions exhibits 31.2% higher power density. The results in this chapter provide a simple yet effective approach to promote the cell performance of nonaqueous redox flow batteries.

Acknowledgements

The work described in this chapter was fully supported by grants from the NSFC, China (No. 51676092 and No. 21676126), Six Talent Peaks Project in Jiangsu Province (2016-XNY-015), and a project funded by the Priority Academic Program Development (PAPD) of Jiangsu Higher Education Institutions, China.

Conflict of interest

There is no conflict of interest to be declared.

Author details

Qian Xu[1]*, Chunzhen Yang[2] and Huaneng Su[1]

1 Institute for Energy Research, Jiangsu University, Zhenjiang, China

2 School of Materials Science and Engineering, Sun Yat-Sen University, Guangzhou, China

*Address all correspondence to: xuqian@ujs.edu.cn

IntechOpen

© 2019 The Author(s). Licensee IntechOpen. This chapter is distributed under the terms of the Creative Commons Attribution License (http://creativecommons.org/licenses/by/3.0), which permits unrestricted use, distribution, and reproduction in any medium, provided the original work is properly cited.

References

[1] Leung P, Li X, De Leon CP, et al. Progress in redox flow batteries, remaining challenges and their applications in energy storage. RSC Advances. 2012;**2**(27):10125-10156. DOI: 10.1039/c2ra21342g

[2] Alotto P, Guarnieri M, Moro F. Redox flow batteries for the storage of renewable energy: A review. Renewable and Sustainable Energy Reviews. 2014;**29**: 325-335. DOI: 10.1016/j.rser.2013.08.001

[3] Soloveichik GL. Flow batteries: Current status and trends. Chemical Reviews. 2015;**115**(20):11533-11558. DOI: 10.1021/cr500720t

[4] Chakrabarti MH, Dryfe RAW, Roberts EPL. Evaluation of electrolytes for redox flow battery applications. Electrochimica Acta. 2007;**52**(5):2189-2195. DOI: 10.1016/j.electacta.2006.08.052

[5] Low CTJ, Walsh FC, Chakrabarti MH, et al. Electrochemical approaches to the production of graphene flakes and their potential applications. Carbon. 2013;**54**(4):1-21. DOI: 10.1016/j.carbon.2012.11.030

[6] Escalante-García IL, Wainright JS, Thompson LT, et al. Performance of a non-aqueous vanadium acetylacetonate prototype redox flow battery: Examination of separators and capacity decay. Journal of the Electrochemical Society. 2015;**162**(3):A363-A372. DOI: 10.1149/2.0471503jes

[7] Leung PK, Sanz L, Flox C, Xu Q, Shah AA, Walsh FC. Recent developments in organic redox flow batteries: A critical review. Journal of Power Sources. 2017;**360**:243-282. DOI: 10.1016/j.jpowsour.2017.05.057

[8] Lewandowski A, Swiderska-Mocek A. ChemInform abstract: Ionic liquids as electrolytes for li-ion batteries: An overview of electrochemical studies. ChemInform. 2010;**41**(38):601-609. DOI: 10.1016/j.jpowsour.2009.06.089

[9] Jhong HR, Wong SH, Wan CC, et al. A novel deep eutectic solvent-based ionic liquid used as electrolyte for dye-sensitized solar cells. Electrochemistry Communications. 2009;**11**(1):209-211. DOI: 10.1016/j.elecom.2008.11.001

[10] Wei D, Ivaska A. Applications of ionic liquids in electrochemical sensors. Analytica Chimica Acta. 2008;**607**(2):126-135. DOI: 10.1016/j.aca.2007.12.011

[11] Lu W, Qu L, Henry K, et al. High performance electrochemical capacitors from aligned carbon nanotube electrodes and ionic liquid electrolytes. Journal of Power Sources. 2009;**189**(2):1270-1277. DOI: 10.1016/j.jpowsour.2009.01.009

[12] Rezaei B, Mallakpour S, Taki M. Application of ionic liquids as an electrolyte additive on the electrochemical behavior of lead acid battery. Journal of Power Sources. 2009;**187**(2):605-612. DOI: 10.1016/j.jpowsour.2008.10.081

[13] Souza RF, Padilha JC, Gonçalves RS, et al. Room temperature dialkylimidazolium ionic liquid-based fuel cells. Electrochemistry Communications. 2003;**5**(8):728-731. DOI: 10.1016/s1388-2481(03)00173-5

[14] Chakrabarti MH, Mjalli FS, AlNashef IM, et al. Prospects of applying ionic liquids and deep eutectic solvents for renewable energy storage by means of redox flow batteries. Renewable and Sustainable Energy Reviews. 2014;**30**:254-270. DOI: 10.1016/j.rser.2013.10.004

[15] Ejigu A, Greatorex-Davies PA, Walsh DA. Room temperature ionic liquid electrolytes for redox flow batteries. Electrochemistry Communications. 2015;**54**:55-59. DOI: 10.1016/j.elecom.2015.01.016

[16] Ding Y, Zhang C, Zhang L, Zhou Y, Yu G. Molecular engineering of organic electroactive materials for redox flow batteries. Chemical Society Reviews. 2018;**47**:69-103. DOI: 10.1039/C7CS00569E

[17] Bahadori L, Hashim MA, Manan NSA, et al. Investigation of ammonium- and phosphonium-based deep eutectic solvents as electrolytes for a non-aqueous all-vanadium redox cell. Journal of the Electrochemical Society. 2016;**163**(5):A632-A638. DOI: 10.1149/2.0261605jes

[18] Miller MA, Wainright JS, Savinell RF. Iron electrodeposition in a deep eutectic solvent for flow batteries. Journal of the Electrochemical Society. 2017;**164**(4):A796-A803. DOI: 10.1149/2.1141704jes

[19] Zhang C, Ding Y, Zhang L, Wang X, Zhao Y, Zhang X, et al. A sustainable redox-flow battery with an aluminum-based deep-eutectic-solvent anolyte. Angewandte Chemie, International Edition. 2017;**56**:7454-7459. DOI: 10.1002/ange.201703399

[20] Zhang L, Zhang C, Ding Y, Ramirez-Meyers K, Yu G. A low-cost and high-energy hybrid iron-aluminum liquid battery achieved by deep eutectic solvents. Joule. 2017;**1**:623-633. DOI: 10.1016/j.joule.2017.08.013

[21] Lloyd D, Vainikka T, Murtomaki L, et al. The kinetics of the Cu^{2+}/Cu^{+} redox couple in deep eutectic solvents. Electrochimica Acta. 2011;**56**(14):4942-4948. DOI: 10.1016/j.electacta.2011.03.133

[22] Lloyd D, Vainikka T, Kontturi K. The development of an all copper hybrid redox flow battery using deep eutectic solvents. Electrochimica Acta. 2013;**100**:18-23. DOI: 10.1016/j.electacta.2013.03.130

[23] Xu Q, Zhao TS, Wei L, Zhang C, Zhou XL. Electrochemical characteristics and transport properties of Fe(II)/Fe(III) redox couple in a non-aqueous reline deep eutectic solvent. Electrochimica Acta. 2015;**154**:462-467. DOI: 10.1016/j.electacta.2014.12.061

[24] Xu Q, Qin LY, Su HN, Xu L, Leung PK, Yang CZ, et al. Electrochemical and transport characteristics of V(II)/V(III) redox couple in a nonaqueous reline deep eutectic solvent: Temperature effect. Journal of Energy Engineering. 2017;**143**(5):04017051. DOI: 10.1016/j.electacta.2014.12.061

[25] Zhang C, Zhang L, Ding Y, Peng S, Guo X, Zhao Y, et al. Progress and prospects of next-generation redox flow batteries. Energy Storage Materials. 2018;**15**:324-350. DOI: 10.1016/j.ensm.2018.06.008

[26] Ding Y, Yu G. The promise of environmentally benign redox flow batteries by molecular engineering. Angewandte Chemie, International Edition. 2017;**56**:8614-8616. DOI: 10.1002/anie.201701254

[27] Zeng S, Zhang X, Gao H, et al. SO_2-induced variations in the viscosity of ionic liquids investigated by in situ fourier transform infrared spectroscopy and simulation calculations. Industrial and Engineering Chemistry Research. 2015;**54**(43):10854-10862. DOI: 10.1021/acs.iecr.5b01807

[28] Li X, Hou M, Han B, et al. Solubility of CO_2 in a choline chloride^{+} urea eutectic mixture. Journal of Chemical and Engineering Data. 2008;**53**(2): 548-550. DOI: 10.1021/je700638u

[29] Leron RB, Li MH. High-pressure volumetric properties of choline chloride-ethylene glycol based deep eutectic solvent and its mixtures with water. Thermochimica Acta. 2012;**546**:54-60. DOI: 10.1016/j.tca.2012.07.024

[30] Leron RB, Li MH. High-pressure density measurements for choline chloride: Urea deep eutectic solvent and its aqueous mixtures at T = (298.15 to 323.15) K and up to 50 MPa. The Journal of Chemical Thermodynamics. 2012;**54**:293-301. DOI: 10.1016/j.jct.2012.05.008

[31] Ali E, Hadj-Kali MK, Mulyono S, et al. Solubility of CO_2, in deep eutectic solvents: Experiments and modelling using the Peng–Robinson equation of state. Chemical Engineering Research and Design. 2014;**92**(10):1898-1906. DOI: 10.1016/j.cherd.2014.02.004

[32] Haghbakhsh R, Raeissi S. Modeling the phase behavior of carbon dioxide solubility in deep eutectic solvents with the cubic plus association equation of state. Journal of Chemical and Engineering Data. 2018;**63**(4):897-906. DOI: 10.1021/acs.jced.7b00472

[33] Wiercigroch E, Szafraniec E, Czamara K, et al. Raman and infrared spectroscopy of carbohydrates: A review. Spectrochimica Acta Part A: Molecular and Biomolecular Spectroscopy. 2017;**185**:317-335. DOI: 10.1016/j.saa.2017.05.045

[34] Xu Q, Qin LY, Su HN, Xu L, Leung P, Yang C. Electrochemical and transport characteristics of V(II)/V(III) redox couple in a nonaqueous deep eutectic solvent: Temperature effect. Journal of Energy Engineering. 2017;**143**:04017051. DOI: 10.1061/(ASCE)EY.1943-7897.0000484

[35] Jiang HR, Zeng YK, Wu MC. A uniformly distributed bismuth nanoparticle-modified carbon cloth electrode for vanadium redox flow batteries. Applied Energy. 2019;**240**:226-235. DOI: 10.1016/j.apenergy.2019.02.051

[36] Shen J, Liu S, He Z, et al. Influence of antimony ions in negative electrolyte on the electrochemical performance of vanadium redox flow batteries. Electrochimica Acta. 2015;**151**:297-305. DOI: 10.1016/j.electacta.2014.11.060

[37] Skyllas-Kazacos M, Chakrabarti MH, Hajimolana SA, Mjalli FS, Saleem M. Progress in flow battery research and development. Journal of the Electrochemical Society. 2015;**158**:R55-R79. DOI: 10.1149/1.3599565

[38] Xu Q, Ji YN, Qin LY, Leung PK, Shah AA, Li YS. Effect of carbon dioxide additive on the characteristics of a deep eutectic solvent (DES) electrolyte for non-aqueous redox flow batteries. Chemical Physics Letters. 2018;**708**:48-53. DOI: 10.1016/j.cplett.2018.07.060

Section 3

Redox and Fish Welfare

Chapter 6

Redox Balance Affects Fish Welfare

Sergio Sánchez-Nuño, Teresa Carbonell and Antoni Ibarz Valls

Abstract

Aquaculture is a growing industry that is increasingly providing a sizable proportion of fishery products for human consumption. Thus, in the last years, several efforts are made in improving fish welfare. As well as in the rest of vertebrates, fish welfare is sensible to a balanced redox status. Numerous inputs like diet and environmental factors could alter this balance. In this sense, the last feeding strategies are focused on developing a more sustainable aquaculture, trying to maintain a redox balance. On the other hand, under culture conditions, animals cannot migrate to more favourable conditions, and environmental stress is one of the most relevant inputs that could compromise redox balance. This chapter is focused on the review of last works in redox balance analysis in Mediterranean aquaculture species and is organized as follows: (1) redox reactions on poikilotherms versus homeotherms; (2) effect of feeding strategies and environmental stress in fish redox balance; and (3) wide vision in fish redox balance.

Keywords: antioxidant enzymes, glutathione redox cycle, lipoperoxidation, protein oxidation, Mediterranean aquaculture

1. Introduction

In the recent decades, the increase in demand of fish products has boosted the development of aquaculture, being today responsible of the major supply of fish for human consumption. Currently, aquaculture is growing faster than other major food production sectors. In 2016, 88%, more than 151 million tons (t) of the fish produced was destined for human consumption (FAO). Thus, the potential of oceans and inland waters could be fundamental for human nutrition in the coming years, due to the expected demographic increase.

Classically, the vast majorly of aquaculture research have been focused on growth studies, feed efficiency, larval maturation or fillet quality. However, as in all intensive animal exploitation, fish culture presents different alterations and pathologies affecting fish welfare, causing great economic losses to the aquaculture sector. Rearing conditions in sea cages expose animals to several stressful inputs like crowding, feed competence, predators and environmental parameters (tides, salinity, temperature, etc.). These stressful conditions affect fish welfare, being more susceptible to infections or even triggering mortality [1, 2], causing important economic loss. To evaluate the individualized effect of each of the inputs and also to test new feeds, aquaculture research applies indoor models to evaluate the cellular and physiological responses. Therefore, a growing interest about fish welfare exists, being the oxidative stress and the fish redox status a new focus of research.

IntechOpen

Like to the studies in mammals' species, the most of fish approaches on redox aquaculture subjects were focused in liver as the main physiological organ. Liver shows powerful enzymatic antioxidant machinery, it is involved on glutathione synthesis and it is a main target of reactive oxygen species (ROS). However, digestive tract, white and red muscle even plasma are of a novel targets for redox studies in fish.

Despite the bibliography reporting redox markers in the last years are increasing, there are only few works tackling on the redox balance and the consequences of its alteration in fish welfare. This chapter aims to evaluate the most relevant and recent studies, and the advances of oxidative status, antioxidant defences and global welfare in fish. Along the chapter we will be focused on the studies which report dietary effects and environmental challenges on fish culture.

2. Redox reactions on ectotherms versus endotherms

2.1 Definition of oxidative stress

Oxygen Free Radicals are highly reactive species which are known to be the major factor in oxidative cell injury via the oxidation and subsequent functional impairment of lipids, carbohydrates, proteins and DNA. In the 1950s, free radicals were first identified in biological systems and were proposed to be involved in pathological processes [3]. The major source of intracellular free radicals is mitochondria due to the presence of an electron transport chain [4, 5], which consumes 85–90% of the oxygen utilized by cell [4, 6]. While passing through the mitochondrial electron transport chain, up to 2% of the total oxygen consumed undergoes one-electron reduction to generate superoxide anion radicals ($O2^{\bullet -}$) and hydrogen peroxide (H_2O_2). This hydrogen peroxide may lead to hydroxyl radical ($OH^{\bullet}$), the most reactive free radical produced in biological systems, with the participation of transition metals in the Haber-Weiss reaction [7]. In addition, different stressors, particularly those induced by environmental physical and chemical factors were reported to increase levels of free radicals. As research on free radicals focused on oxygen radicals, with some other forms of non-radical active oxygen, they are collectively referred to as reactive oxygen species (ROS).

As the formation of reactive oxygen species (ROS) is a part of natural cellular oxidative metabolism, the question if living organisms possess regulated enzymatic systems to defend against ROS, suddenly arises. This was first confirmed by McCord and Fridovich [8] who discovered the enzyme superoxide dismutase (SOD) and demonstrated that living organisms have developed protective mechanisms against ROS. Over time, this was supported by continuing discoveries of several mechanisms by which ROS can be neutralized: antioxidant enzymes and low molecular mass antioxidants. During normal oxidative metabolism ROS are produced continually, but they are scavenged by superoxide dismutase (SOD), glutathione peroxidase and catalase [9]. Other small molecular antioxidants: glutathione, ascorbic acid and a-tocopherol are also involved in the detoxification of free radicals. Those reported evidences lead to Helmut Sies [10] to first propose the following definition of oxidative stress as 'imbalance between oxidants and antioxidants in favour of the oxidants, potentially leading to damage'. Oxidative stress was considered to be harmful, while antioxidants provided defence and prevention of tissue damage. However, ROS were recently found to play signalling roles not only in ROS-related processes, but in many basic functions such as fertilization, growth, and differentiation [11–14].

2.2 Redox balance

We have described how reactive oxygen species (ROS) can be both harmful and beneficial. Consequently, under physiological conditions, the cellular redox equilibrium is tightly regulated on the one hand by pro-oxidants and on the other by enzymatic and non-enzymatic antioxidants (**Figure 1**). Due to the central role of ROS in many pathologies, restoring the redox balance forms an innovative target in the development of new strategies for treating several conditions. For example, Coenzyme-Q and its redox status -that was mostly found in the reduced form- have been proposed as an adaptation to different thermal environments in Antarctic fishes [15], and to reflects species-specific ecological habits and physiological constraint associated with oxygen demand represent an adaptation to environmental oxygen availability in coral reef fishes [16]. Recently, the determination of the redox balance of liver Coenzyme-Q from fish has been observed to have a potential, based on physiological principles, to be used as a practical biomarker for polycyclic aromatic hydrocarbon (PAH) contamination in aquatic biotopes [17].

The main marker of oxidative damage in fish tissues have been considered as lipid peroxidation (LPO, usually measured as ThioBarbituric Reactive Substances, TBARS) [18–20]. However, the application of mammals' techniques to measure the direct oxidation of the amino acid side chains, so-called the Advanced Oxidation Protein Products, (AOPPs) or/and the tissue accumulation of 4-hydroxinonenal (4-HNE) cross-linked proteins are novel approach in the study of oxidative insult in fish species [21].

To cope with the oxidative damage resulting from metabolism, animals use non-enzymatic defences, such as thiol groups and glutathione, and enzymes with antioxidant activity [1, 20, 22]. Thus, in fish as in mammals, antioxidant liver capacity has been classically measured via total glutathione, (tGSH) and its oxidized and reduced forms (GSH and GSSG, respectively), and via the main antioxidant enzyme activities: superoxide dismutase (SOD), catalase (CAT), glutathione peroxidase (GPX) and glutathione reductase (GR).

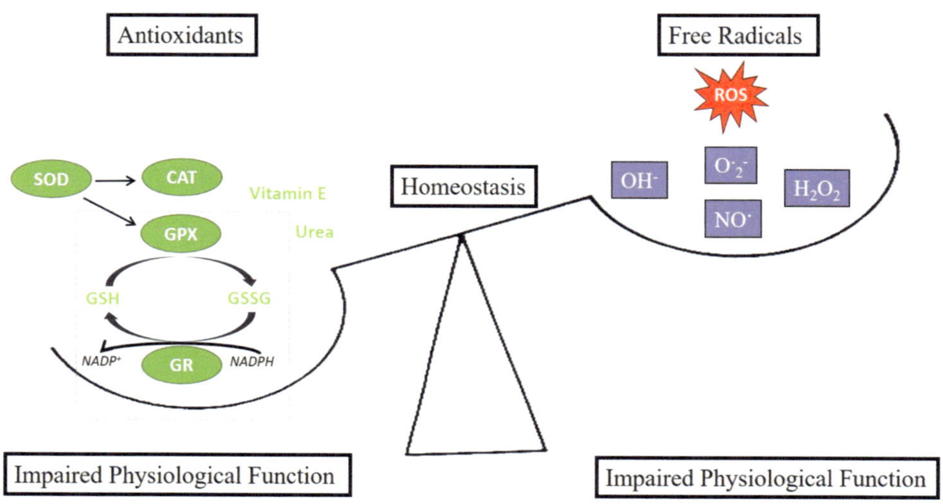

Figure 1.
Homeostasis in redox balance.

2.3 Comparative studies: ectotherms versus endotherms

The generic term 'fish' comprises an extremely diverse group of vertebrates, which represents 40% of the world's vertebrates. Fish are adapted to aquatic environment, showing a wide type of adaptations to different environmental conditions of temperature, salinity, oxygen level and water chemistry. To survive in such diverse environments, fish need high adaptive potential. If it is overwhelmed, organisms may enter stress conditions. The most studied environmental stress conditions in aquatic environments include changes in salinity, ion composition, and temperature and oxygen availability. Recently, they also included pollutants exposure due to human activity.

If the previous studies of comparative physiology were focused on the fish availability of O_2, the generation of ROS and its harmful consequences are currently taken into account. In fishes, the main physiological source of ROS is also the mitochondria, and the mechanism of ROS production similar to that of mammal. However, the fraction of total mitochondrial electron flux that generate H_2O_2 (the fractional electron leak), was far lower in rat than in all the ectothermic fishes assayed [23]. Results previously published concluded that mitochondria of true endotherms (such as birds and mammals) produce lower rates of ROS generation compared to fish and showed higher levels of antioxidant enzymes compared to fish [24].

Despite similarities in mitochondrial coupling mechanisms and proton leakage, differences in mitochondrial function between mammals and fish have been reported. A greater phospholipid unsaturation, the presence of cardiolipin and the absence of cholesterol in the mitochondrial membranes of fish can establish different aspects of mitochondrial functionality in fish. In fact, the presence of the phospholipid cardiolipine and the absence of cholesterol may confer a high structural flexibility in fish mitochondria. Moreover, differences in membrane phospholipid composition involving polyunsaturated fatty acids (PUFA) may play a role in the proton leak across the mitochondrial membranes (which can explain greater proton leak in endoherms when compared to ectotherms) and in the lipid peroxidation process. It has been previously demonstrated the correlation between the consumption of polyunsaturated fatty acid diet and the enhancement of endogenous lipid peroxidation, in rats [1, 25, 26]. As the increased unsaturation of fatty acids leads to increased parameters of oxidative stress damage, this increases the oxidative challenge to fish tissues. However, some important differences in the lipid composition of membranes have been reported for phylogenetic distinct marine fish species [27]. In the referred work, the elasmobranch *Raja erinacea* showed lower percentage of polyunsaturated fatty acids when compared to non-elasmobranch species. Fishes of the Antarctic seas showed relatively high proportion of polyunsaturated fatty acids in mitochondria [28], which make them prone to lipid peroxidation. Indeed, reported TBARS levels are much higher in fish liver then in mammals [24].

As reported above, all aerobic forms of life developed antioxidant defences. Both, low-molecular weight antioxidants and antioxidant enzymes like catalase, superoxide dismutase and glutathione- dependent enzymes, have been detected in different fish species [1]. Many studies confirm that enzymatic antioxidant activities in fish were lower than in endotherms. As fish refers to different evolved species, some studies have been performed to determine whether antioxidants correlate with phylogenetic position [29]. Marine fishes showed high levels of the antioxidant vitamin E [30] and elasmobranchs compensate limited antioxidant enzymes with high levels of glutathione and urea [31]. It seems that low molecular weight antioxidants appeared early in evolution to later develop enzymatic antioxidant systems.

3. Advances in redox balance of temperate fish species

3.1 Redox balance in feeding strategies

Usually, the main aim of the diet trials in fish pursue to improve growth performance as well as food conversion and efficiency, guaranteeing a high-quality product, reducing fish mortality and with economic positive gains. Recently, feeding studies include redox balance markers as additional indicators of the fish physiological status. **Table 1** represents the most relevant feeding strategies in Mediterranean fish related to redox balance.

	Tissue	Diet strategy	Oxidants	Antioxidants
Diet effect				
Dicentrarchus labrax	Liver, intestine	Soy based diets and Taurine [40]	LPO	SOD, CAT
	Liver, intestine	Substitution diets and carbohydrates reduction [32]	LPO	SOD, CAT, GR, GPX, G6PDH, GSH and GSSG
Sparus aurata	Liver	Supplemented diet Met and Tea [41]	LPO	SOD, CAT, GR, GPX, G6PDH, GSH and GSSG
	Mucus, gut, skin	Palm fruit extract and probiotics⁺ [43]	LPO, H_2O_2	SOD, CAT, GR
	Intestine	Sprayed diet porcine protein [42]	TBARS	CAT, GR, GST, GPX
	Liver, intestine	Soybean and wheat with arginine [39]	LPO	SOD, CAT, GR, GPX, G6PDH
	Liver, intestine	Soybean and wheat with glutamine [38]	LPO	SOD, CAT, GR, GPX, G6PDH
	Liver, intestine	Substitution diets and carbohydrates reduction [34]	LPO	SOD, CAT, GR, GPX, G6PDH, GSH and GSSG
	Liver	Soybean replacement [36]		SOD, CAT, GR, GPX, GST
	Liver	Soybean replacement supplemented with Met and phosphate [37]		SOD, CAT, GR, GPX, GST
Dentex dentex	Liver, white muscle	Different carbohydrates and carbohydrates amount [35]	LPO, carbonylated proteins	SOD, CAT, GR, GPX, G6PDH
Diet to enhance welfare				
Dicentrarchus labrax	Liver, intestine	Substitution diets and carbohydrates reduction* [33]	LPO	SOD, CAT, GR, GPX, G6PDH
Solea senegalensis	Liver	Reduced dietary proteins** [19]	LPO	SOD, CAT, GR, GPX, G6PDH
Sparus aurata	Liver, plasma	Lipid reduction** [21]	LPO, AOPP	SOD, CAT, GR, GPX, GSH and GSSG
Scophthalmus maximus	Liver	Prebiotics** [53]	LPO	SOD, CAT, GR, GPX, G6PDH

⁺*Gene expression of antioxidant enzymes.*
**To enhance the response to handling stress.*
***To enhance the response to thermal stress.*

Table 1.
Last decade works related to feeding strategies in Mediterranean teleost fish.

3.1.1 Fish oil reduction

During the last two decades several feeding strategies and economic needs have been promoted the reduction of total protein and the substitution of both fish meal and fish oil for other sources such as vegetable, algae and even insects or yeasts. However, a common strategy for the sector cannot be achieved due to the great variability within the biology of fish species concerning, for instance marine versus freshwaters; carnivorous versus. herbivorous and omnivorous; temperate versus eurytherms, etc.

The replacement of fish oil (FO) by vegetal oil blend (VO) (20% rapeseed, 50% linseed, and 30% palm oils) would promotes lower levels of lipoperoxidation (LPO) products in liver and intestine, together with higher enzymatic activities of GPX and GR in intestine of sea bass, *Dicentrarchus labrax* [32, 33]. When a stressful condition introduced, carbohydrate rich diets also diminished LPO products and increased GR and glucose 6-phosphodihydrogenase (G6PDH) and VO diets enhanced GPX and G6PDH activities [33]. The same diet strategy was conducted in other species such as the gilthead sea bream, *Sparus aurata*, evidencing that the enriched diets with starch carbohydrates promoted the antioxidant defences by reducing oxidized glutathione and lower LPO products [34]. By its way, in common dentex, *Dentex dentex*, higher carbohydrates levels inclusion in diets, increased GPX in liver and white muscle and GR in liver and 18 and 24% and decreased oxidative products as protein carbonylation in liver and LPO in liver and white muscle [35].

In addition to the studies of reducing FO consumption by reducing from VO substitution, some authors had considered that the lipid content of commercial diets must be lowered. Sánchez-Nuño et al. [21] evidenced that the reduction from 18–14% in the lipid content of the diet did not affect *Sparus aurata* growth, glutathione levels or enzyme activities, but did reduce the amount of LPO.

3.1.2 Fish sources reduction in diet formulation

The substitution (total and partial) of fish meal proteins by vegetable sources (soybean and wheat proteins as the main used products in the last decades) seems to be another relevant topic in diet substitution. Although a classic discrepancy exists on the benefits to substitute protein by carbohydrate and its effects on fish growth, for several fish species the carbohydrate inclusion to replace protein would benefit redox status, and if the growth performance is not affected, this strategy could be considered as beneficious for fish welfare.

In gilthead sea bream [36] fed with formulated diets replacing fish meal by soybean protein at 20, 40 and 60% showed a gradually increase of liver antioxidant enzymes activities (SOD, CAT, GPX, GR) according to higher levels of fish meal replacement. Despite any oxidized products were not evaluated, growth performance and immunity markers were negatively affected, suggesting that fish proteins are essential in diet formulation. However, when soybean diets were supplemented with methionine and phosphate fish redox status enhanced significantly [37]. In the same line, glutamine and arginine supplementation would improve deleterious effects of higher fish meal substitutions [38, 39] although with lower benefits observed with methionine inclusion. Taurine supplementation was also proposed to improve fish welfare when fish meal is replaced by vegetable oils but with not clear benefits [40].

Irrespective to sources substitution several studies approached specific supplements to improve the redox balance to benefit fish welfare. A combination of methionine and white tea dry leaves supplementation to a commercial diet were proposed in gilthead sea bream by [41]. After 4 weeks feeding this diet liver SOD and CAT activities were increased, although no reduced LPO levels or higher glutathione amount were evident.

Other novel techniques and feed strategies were also assayed in fish nourishment studies. For instance, the inclusion of spray-dried plasma from porcine blood (SDPP) has been evaluated because of in mammals evidenced great results in immune and redox status. In fish fed with 3% of SDPP showed lower CAT, GR, and Glutathione S-Transferase (GST) enzymatic activity [42]. Other emerging strategy is to enrich diets with bacterial probiotics and to study its effects also in redox status. A recent study including *Shewanella putrefaciens*, *Pdp11* and *Bacillus* sp. evaluated the gene expression of redox balance markers in gill, intestine and epidermal mucosae [43]. Experimental diets alter the expression of the studied antioxidant genes, primarily in the gill and skin. Furthermore, the tested probiotics and mainly, the palm fruits extracts had significant antioxidant properties especially after feeding for 30 days.

3.2 Redox balance in response to environmental stress conditions

Another of the most relevant aspects in animal culture is the exposure to continuous environmental stressors and, in the most cases, fish are challenged to various types of abiotic or biotic stressors simultaneously. Whereas in the wildlife, fish have the freedom to migrate to locations where environmental conditions are within their tolerance range or to escape from disfavour ones, in culture conditions their confinement makes the escape impossible and entails the need to face up stressors with physiological responses. The classical abiotic and biotic stressors for fish culture are infections, parasites, changes in water salinity, exposure to dissolved heavy metals, the decrease in oxygen availability, the food access limitation, human handling, higher densities and mainly, at temperate latitudes, the natural and seasonal variations in water temperature. From the last two decades, the implications of physiological redox balance to face up culture conditions stressors are of the main interest for scientists and farmers. Some of the most relevant works are summarized in **Table 2** and discussed below.

	Tissue	Environmental stress	Oxidants	Antioxidants
Starvation				
Dicentrarchus labrax	Liver, intestine, white and red muscle	Starvation 1–2 months [44]	LPO	SOD, CAT, GPX
Handling				
Dicentrarchus labrax	Liver, intestine	Handling stress [33]	LPO	SOD, CAT, GR, GPX, G6PDH
Salinity				
Dicentrarchus labrax	Liver	Salinity and ammonia toxicity [45]	LPO, H_2O_2	SOD, CAT, GR, GPX, GSH and GSSG
Hypoxia				
Sparus aurata	Liver, heart	Hypoxia [47]	LPO	CAT, GR, GPX, GST
Temperature				
Dicentrarchus labrax	White muscle	18 versus 24 versus 28°C [52]	LPO	CAT
Solea senegalensis	Liver	12 versus 18°C [19]	LPO	SOD, CAT, GR, GPX, G6PDH

	Tissue	Environmental stress	Oxidants	Antioxidants
Sparus aurata	Liver	8 versus 20°C [50]	LPO NO	Liver proteome
	Liver, heart, white and red muscle	18 versus 24 versus 28°C and CO2 [48]	LPO	SOD, CAT, GR
	Liver, heart, white and red muscle	Seasonal fluctuation [51]	LPO	SOD, CAT, GPX, XO
	Liver, plasma	14 versus 22°C [21]	LPO, AOPP	SOD, CAT, GR, GPX, GSH and GSSG
Scophthalmus maximus	Liver	15 versus 20°C [53]	LPO	SOD, CAT, GR, GPX, G6PDH

Table 2.
Last decade works related to environmental stress in Mediterranean teleost fish.

3.2.1 Food deprivation

Food deprivation in wild fish, is a common fact and the physiological impact tends to be less aggressive than in higher vertebrates. Starvation reduces energy metabolism and cellular activity, resulting to lower oxygen consumption which can lead to oxidative stress caused by hypoxia [44]. Due to the economical repercussion for the industry, fasting is widely described for the vast majority of fish, classically focusing in metabolism and growth. However, few studies approach to redox balance and on the consequences of an unbalanced oxidative context. For instance, long-term starvation (1–2 months) in sea bass, *Dicentrarchus labrax* altered the redox balance in red muscle, white muscle, intestine and liver. Moreover, refeeding period has been demonstrated as a crucial and additional stressful period [44]. Alterations in redox balance seem to be tissue dependent, but triggering LPO in liver. However, food deprivation and refeeding enhanced antioxidant enzyme activities in intestine, and decreased SOD activity in red and white muscle, and CAT and GPX in white muscle.

3.2.2 Salinity

In seawater, salinity variations influence physiological processes and condition fish species abundance, being in consequence, one of the most determinant environmental stressors. Some species are tolerant with variable salinities, euryhaline species, and could be related with its migratory behaviour. Thus, the physiological adaptative strategies to cope with the changing environmental salinities were very recurrent on fish biology studies. Recent studies suggested that changes in salinity may also induce oxidative stress compromising antioxidant defences. However, it becomes difficult to extrapolate a solution under culture conditions after evaluating the response to different stressors independently. For that reason, the current trend in salinity studies is to combine different stressors to evaluate the joint response. For instance, [45] studied the combined effects on the sea bass redox balance of the salinity and the High Environmental Ammonia (HEA) The results evidenced the antioxidant defences strength in this specie in low-saline seawaters (up to 10 ppt) remaining unaltered even with increased HEA levels. However, it seems to be a limit at hypo-saline environment (2.5 ppt) in combination with HEA exposure where antioxidant defence were compromised.

3.2.3 Hypoxia

As it was referred above hypoxia is one of the main factors resulting in an oxidative attack in fish as in mammals. However, the mechanisms of hypoxia-induced oxidative stress remain still unclear. One of the hypotheses supports that a reduction in the mitochondrial electron-transport chain efficiency may contribute to ROS generation [46]. Sustained severe hypoxic conditions could trigger fish mortality. As an example, [47] demonstrated in *Sparus aurata* an increase of LPO, increased CAT activity and reduced GR and GPX hepatic enzymes in response to environmental O_2 fluctuations after an hypoxia exposure. All these effects seem to be mitigated by the dietary supplementation of seaweeds, suggesting its protective role against oxidative stress. The principal causes of hypoxia in the sea are crowding and the increase of water temperature, being the climate change one of the main effectors. Because climate change is currently a trending topic, some researchers have focused their efforts on finding new strategies to mitigate the effects of high temperatures [48].

3.2.4 Temperature

One of the most relevant environmental inputs is thermal variations, being a challenge for poikilothermic animals. When the seawater temperature escapes from the limits of intraspecific tolerance, wild animals can respond in different ways, being the physiological escape (migration) one of the most common responses [2]. Under culture conditions, animals are not able to migrate and are obligated to face these temperatures, forcing an adaptation and implying physiological changes. Focusing on the physiological 'symptoms', loss of appetite would be the first response to stress due to low temperatures [49]. Due to the productive interest and the evidences of the increasingly extreme seasonal temperatures, great efforts have been devoted to improving animal welfare during the thermal changes, being the influence on metabolism and redox balance analysis key topics. In last decade, several studies have introduced novel techniques in fish to evaluate the redox balance. Ibarz et al. [50] approached proteomics to evaluate the cold exposure in sea bream liver. Their results demonstrated that after 10 days at 8°C LPO increased by 50% and antioxidant proteins such as betaine-homocysteine-methyl transferase (BHMT, related with glutathione synthesis), GST and CAT were downregulated, suggesting that this species are very sensitive by low temperatures. The warming response was also evaluated after 10 days at 24 and 28°C, evidencing that growing temperatures also increase LPO and stimulates CAT and GR activities in several tissues like heart, muscle (white and red) and liver [48]. In addition to acute exposure to thermal stress, some studies have focused on the medium-long-term effect of temperature challenge, trying to understand the physiological response. In this way, Sánchez-Nuño et al. [21] described the redox balance behaviour (in liver and in plasma) after 50 days of cold exposure at 14°C. Cold exposure compromised antioxidant enzyme activities mainly CAT and GR, which subsequently affected the glutathione redox cycle and caused an acute reduction in total hepatic glutathione levels. During temperature recovery, antioxidant enzymatic machinery was gradually restored but the glutathione redox cycle was not recovered. Despite field studies are not very common, it was evaluated the effect of seasonal temperature fluctuation in heart, liver and muscle to understand the adapted physiological state, including redox balance [51]. Results evidenced clear seasonal metabolic patterns involving oxidative stress during summer as well as winter, but more prominent during warming because of the increased aerobic metabolism. During cold acclimatization and under increased temperatures LPO was higher.

By its way, [52] evaluated in sea bass white muscle the LPO and CAT activity when increasing temperature from 16–18, 24 and 28°C and maintained during 15 and 30 days, describing that temperature rise increase LPO and CAT activity. The exposure time conditioned the response, evidencing an acclimation after 30 days at

28°C. Moreover, it was evaluated the effect of rearing water temperature in turbot juveniles, *Scophthalmus maximus* (15 days at normal 15°C and higher 20°C), evidencing lower CAT and GPX enzyme activities [53]. In Senegalese sole, *Solea senegalensis,* cold exposure at 12°C increased GR activity and LPO while decreased SOD activity in liver [19]. Moreover, in this specie decreasing protein content in diet from 55–45% worsened redox balance. All these recent studies reinforce the idea that temperature challenges strongly compromise redox balance in fish temperate species.

4. A wide vision of redox balance in fish: looking for new markers

Fish, like all other organisms, must have a balance between the production of oxidative substances (ROS and RNS) and antioxidant defences, and are of particular interest as they experience a multitude of above-mentioned stressors. To protect themselves against the potentially highly damaging oxidants, organisms have evolved a system to either prevent or repair the effects of oxidative stress. Prevention comes in the form of antioxidants, which can either be enzymatic (SOD, CAT, GPX, GR) or non-enzymatic molecules (mainly glutathione and vitamins C and E), carotenoids and other small molecules. These are the most commonly measured oxidative markers and antioxidants in fish biology (recently revised [54]). However, recent advances on redox studies in mammals suggest other markers to be considered widening to associated pathways of redox balance. Some of them should be also considered when studying redox balance in fish. **Figure 2** attempts to provide this broad range of markers also in fish: including markers of protein oxidation levels, metabolic enzymes related to glutathione synthesis, repair/refolding of oxidized proteins or main protein degradation processes (via ubiquitin-proteasome system, UPS, or lysosomal proteolytic fate).

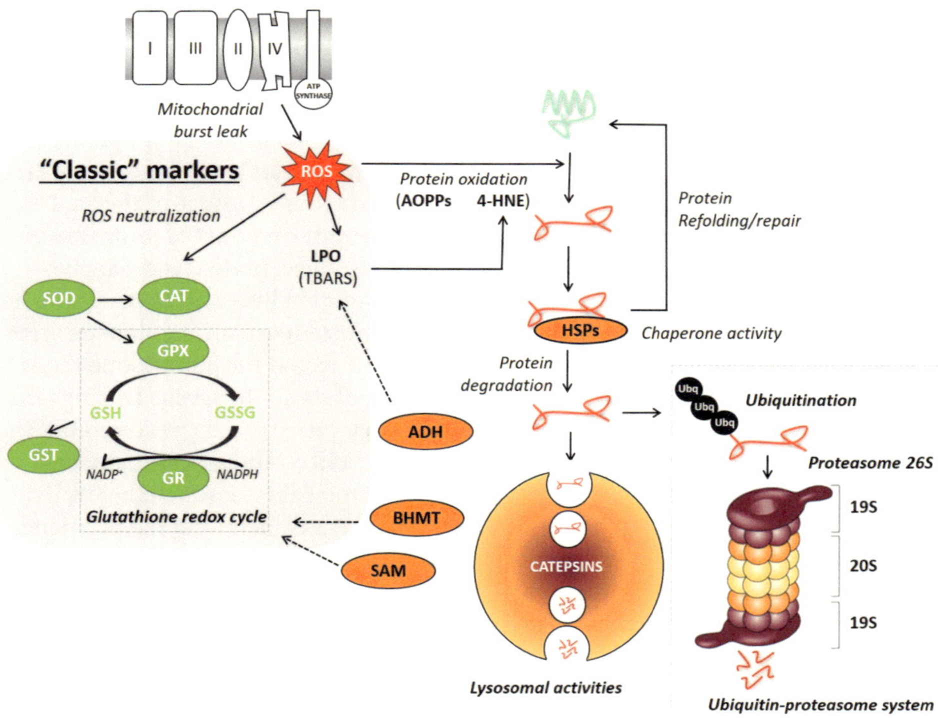

Figure 2.
'Wide view' of redox balance markers.

Beyond the LPO measurements (via TBARS analyses), it was recently reported that advanced oxidation protein products (AOPPs) could act as markers of oxidative stress, resulted from the direct oxidation of the amino acid side chains of proteins as a consequence of metal toxicity in freshwater species [55, 56] or produced at low temperatures in marine species [21]. Another marker recently used in mammals is the accumulation of conjugated proteins with 4-HNE as the major biomarker relating lipoperoxidation and protein oxidation [57, 58]. 4-HNE is the most toxic product of lipid peroxidation reported in mammals damaging proteins by adding covalent adducts and accelerating protein aggregation [59, 60]. Although no data exists in fish on HNE products analyses, a preliminary study in gilthead sea bream would demonstrate the existence of a singular pattern of 4-HNE oxidized proteins in liver with the proteins ranging 40–50 and 75–100 kDa the main target of lipid peroxidation (own data unpublished).

As proteins are the major constituents of cellular organization and metabolism, effects of oxidative attack on protein structure, stability and folding deserve careful consideration in fish species. Several cellular pathways exist which repair and eliminate damaged proteins and thus prevent their accumulation and aggregation. One of the first mechanisms to cope with protein damage is binding with chaperones or 'heat shock proteins' (HSPs). Thus, a few studies recently addressed the changes on HSPs levels/expression in fish when the redox balance is challenged. For instance, it was reported on several fish tissues the expression of HSP70 and HSP90 protein expression levels in response to temperature seasonal challenges [61, 62]. When a protein is irremediably damaged, its fate is to be recycled via UPS degradation pathways, or to be removed/autophaged via a lysosomal degradation process. Protein degradation via UPS involves two discrete and successive steps: tagging of the substrate protein by the covalent attachment of multiple ubiquitin molecules, and the subsequent degradation of the tagged protein by the 26S proteasome, composed of the catalytic 20S core and the 19S regulator [63]. The capacity to remove damaged proteins by the proteasome may prevent oxidative stress and it has been suggested that this is also part of the antioxidant defences in mammals [64]. In fish which inhabiting permanently temperature-fluctuant aquatic environments, this enzymatic complex could play a key role in antioxidant defence systems [65]. The analysis of UPS system markers is still scarce or null in fish and to obtain a profile of protein-ubiquitination labelling according the MW or the analyses of main proteasome subunits (catalytic 20S core or 19S regulatory) should be of further interest on fish studies related to redox balance.

The aggregation of proteins as a result of the accumulation of cross-linked 4-HNE proteins cannot be degraded through the UPS even could block its correct functioning. Then, lysosome acts to eliminate protein aggregates. The lysosomal system has an elevated non-selective protein degradation capacity as a result of the combined random and limited action of various proteases, with the cathepsin family being one of the most important in mammals [59, 66] and also in fish [67]. Thus, as the fate of 4-HNE protein conjugates is preferably the lysosomal degradation pathway, the cathepsin activities should be another target on the study of redox balance.

Finally, we encourage fish biology researcher to study other enzymes which their activities are also strongly related with antioxidant defence. Both TBARS and HNE are aldehydic products, relatively stable and capable of roaming freely and attacking molecules, e.g. DNA, proteins, lipids far from their origin [68]. Thus, aldehyde dehydrogenase (ADH) is an enzyme involved in the oxidation pathway, and is complementary to the GST pathway, which reduces the potential damage of these peroxide products by oxidizing them and removing them from inside the cells. As the glutathione plays a central role in lipid peroxide detoxification, reducing the peroxides to their corresponding alcohols, the study of enzymes involved in glutathione synthesis seems adequate to evaluate cell redox potential and response. The BHMT and the adenosine-methionine synthetase (SAM) are enzymes involved in antioxidant mechanisms through the synthesis of S-adenosylmethionine and through maintain its steady-state levels which is a crucial component of methylation reactions and a biosynthetic precursor of glutathione. Under acute cold stress the expression of these enzymes of glutathione synthesis as well as ADH are affected in gilthead sea bream [50].

Overall, we hope that the proposed markers in **Figure 2** from that 'wide view' on redox balance and related processes in fish. Can contribute to expanding knowledge of the relevant oxidant products (LPO, AOPPs, HNE), the classic enzymes studied (SOD, CAT, GPX, GR, GST) and the associated processes affected such as protein repairing/protection machinery and protein turnover by the oxidant insult.

5. Conclusions

In the last years, several efforts have been made in the Aquaculture industry to improve fish welfare. Changes in environmental conditions and diets can alter the redox balance, thus affecting the welfare state. Most redox studies in fish focuses on classical markers such enzymatic and non-enzymatic antioxidants, and lipoperoxidation products as oxidative damage markers. In this paper we present evidence that the study should be extended to the associated pathways of redox balance including: markers of protein oxidation levels, metabolic enzymes related to glutathione synthesis, repair/refolding of oxidized proteins or main protein degradation processes.

Author details

Sergio Sánchez-Nuño, Teresa Carbonell and Antoni Ibarz Valls*
Department of Cellular Biology, Physiology and Immunology, Faculty of Biology, Universitat de Barcelona, Barcelona, Spain

*Address all correspondence to: tibarz@ub.edu

IntechOpen

© 2019 The Author(s). Licensee IntechOpen. This chapter is distributed under the terms of the Creative Commons Attribution License (http://creativecommons.org/licenses/by/3.0), which permits unrestricted use, distribution, and reproduction in any medium, provided the original work is properly cited.

References

[1] Martínez-Álvarez RM, Morales AE, Sanz A. Antioxidant defenses in fish: Biotic and abiotic factors. Reviews in Fish Biology and Fisheries. 2005;**15**(1-2):75-88

[2] Donaldson MR, Cooke SJ, Patterson DA, Macdonald JS. Cold shock and fish. Journal of Fish Biology. 2008;**73**(7):1491-1530

[3] Gerschman R, Gilbert DL, Nye SW, Dwyer P, Fenn WO. Oxygen poisoning and X-irradiation: A mechanism in common. Science. 1954;**119**(3097):623-626

[4] Boveris A, Chance B. The mitochondrial generation of hydrogen peroxide. General properties and effect of hyperbaric oxygen. Biochemical Journal [Internet]. 1973;**134**(3):707-716. Available from: http://www.biochemj.org.sire.ub.edu/content/134/3/707.abstract [Cited: 10 May 2019]

[5] Halliwell B, Gutteridge JMC. Free Radicals in Biology and Medicine [Internet]. 5th ed. Oxford University Press; 2015. Available from: https://books.google.es/books?hl=es&lr=&id=3DlKCgAAQBAJ&oi=fnd&pg=PP1&dq=Free+Radicals+in+Biology+and+Medicine.&ots=bomH9SDpkY&sig=UL_E9pO5VPiMCXYvSfvqbgbn0Rc#v=onepage&q=Free Radicals in Biology and Medicine.&f=false [Cited: 10 May 2019]

[6] Loschen G, Azzi A, Richter C, Flohé L. Superoxide radicals as precursors of mitochondrial hydrogen peroxide. FEBS Letters [Internet]. 1974;**42**(1):68-72. Available from: http://doi.wiley.com/10.1016/0014-5793%2874%2980281-4 [Cited: 10 May 2019]

[7] Gutteridge JMC, Halliwell B. Free radicals and antioxidants in the year 2000: A historical look to the future. Annals of the New York Academy of Sciences [Internet]. 2000;**899**(1):136-147. DOI: 10.1111/j.1749-6632.2000.tb06182.x

[8] McCord JM, Fridovich I. The utility of superoxide dismutase in studying free racical reactions. The Journal of Biological Chemistry. 1969;**244**(November 25):6056-6063

[9] Yu BP. Cellular defenses against damage from reactive oxygen species. Physiological Reviews [Internet]. 1994;**74**(1):139-162. Available from: http://www.ncbi.nlm.nih.gov/pubmed/8295932 [Cited: 10 May 2019]

[10] Sies H. Brüne B. Oxygen Biology and Hypoxia: Academic Press; 2007. 543p

[11] Christman MF, Morgan RW, Jacobson FS, Ames BN. Positive control of a regulon for defenses against oxidative stress and some heat-shock proteins in *Salmonella typhimurium*. Cell [Internet]. 1985;**41**(3):753-762. Available from: https://www-sciencedirect-com.sire.ub.edu/science/article/pii/S0092867485800568 [Cited: 10 May 2019]

[12] Tartaglia LA, Storz G, Ames BN. Identification and molecular analysis of oxyR-regulated promoters important for the bacterial adaptation to oxidative stress. Journal of Molecular Biology [Internet]. 1989;**210**(4):709-719. Available from: https://www-sciencedirect-com.sire.ub.edu/science/article/pii/0022283689901046 [Cited: 10 May 2019]

[13] Scandalios JG. Oxidative stress: Molecular perception and transduction of signals triggering antioxidant gene defenses. Brazilian Journal of Medical and Biological Research [Internet]. 2005;**38**(7):995-1014. Available from: http://www.scielo.br/scielo.php?script=sci_arttext&pid=S0100-879X2005000700003&lng=en&tlng=en [Cited: 10 May 2019]

[14] Semchyshyn H, Bagnyukova T, Storey K, Lushchak V. Hydrogen peroxide increases the activities of regulon enzymes and the levels of oxidized proteins and lipids. Cell Biology International [Internet]. 2005;**29**(11):898-902. DOI: 10.1016/j.cellbi.2005.08.002

[15] Giardina B, Gozzo ML, Zappacosta B, Colacicco L, Callà C, Mordente A, et al. Coenzyme Q homologs and trace elements content of Antarctic fishes *Chionodraco hamatus* and *Pagothenia bernacchii* compared with the mediterranean fish *Mugil cephalus*. Comparative Biochemistry and Physiology - Part A [Internet]. 1997;**118**(4):977-980. Available from: https://www-sciencedirect-com.sire.ub.edu/science/article/pii/S0300962997867850 [Cited: 10 May 2019]

[16] Gagliano M, Dunlap WC, de Nys R, Depczynski M. Ockham's razor gone blunt: Coenzyme Q adaptation and redox balance in tropical reef fishes. Biology Letters [Internet]. 2009;**5**(3):360-363. Available from: http://www.royalsocietypublishing.org/doi/10.1098/rsbl.2009.0004 [Cited: 10 May 2019]

[17] Hasbi G, de Nys R, Burns K, Whalan S, Dunlap WC. Hepatic coenzyme Q redox balance of fishes as a potential bioindicator of environmental contamination by polycyclic aromatic hydrocarbons. Biology Letters [Internet]. 2011;**7**(1):123-126. Available from: http://www.royalsocietypublishing.org/doi/10.1098/rsbl.2010.0600 [Cited: 10 May 2019]

[18] Tocher DR, Mourente G, Van Der Eecken A, Evjemo JO, Diaz E, Bell JG, et al. Effects of dietary vitamin E on antioxidant defence mechanisms of juvenile turbot (*Scophthalmus maximus* L.), halibut (*Hippoglossus hippoglossus* L.) and sea bream (*Sparus aurata* L.). Aquaculture Nutrition. 2002;**8**(September 2001):195-207

[19] Castro C, Pérez-Jiménez A, Guerreiro I, Peres H, Castro-Cunha M, Oliva-Teles A. Effects of temperature and dietary protein level on hepatic oxidative status of Senegalese sole juveniles (*Solea senegalensis*). Comparative Biochemistry and Physiology - Part A [Internet]. 2012;**163**(3-4):372-378. DOI: 10.1016/j.cbpa.2012.07.003

[20] Lushchak VI. Contaminant-induced oxidative stress in fish: A mechanistic approach. Fish Physiology and Biochemistry. 2016;**42**(2):711-747

[21] Sánchez-Nuño S, Sanahuja I, Ferández-Alacid L, Ordóñez-Grande B, Fontanillas R, Fernández-Borràs J, et al. Redox challenge in a cultured temperate marine species during low temperature and temperature recovery. Frontiers in Physiology [Internet]. 2018;**9**:923. Available from: https://www.frontiersin.org/articles/10.3389/fphys.2018.00923/abstract [Cited: 13 July 2018]

[22] Srikanth K, Pereira E, Duarte AC, Ahmad I. Glutathione and its dependent enzymes' modulatory responses to toxic metals and metalloids in fish—A review. Environmental Science and Pollution Research. 2013;**20**(4):2133-2149

[23] Wiens L, Banh S, Sotiri E, Jastroch M, Block BA, Brand MD, et al. Comparison of mitochondrial reactive oxygen species production of ectothermic and endothermic fish muscle. Frontiers in Physiology [Internet]. 2017;**8**:704. Available from: http://journal.frontiersin.org/article/10.3389/fphys.2017.00704/full [Cited: 10 May 2019]

[24] Wilhelm Filho D, Marcon JL, Fraga CG, Boveris A. Antioxidant defenses in vertebrates: Emphasis on fish and mammals. Trends in Comparative Biochemistry & Physiology. 2000;7:33-45

[25] Mitjavila MT, Carbonell T, Puig-Parellada P, Cambon-Gros C,

Fernandez Y, Pipy B, et al. Combined effect of an essential fatty acid-deficient diet and iron levels on lipid peroxidation in inflamed rats. Food Additives & Contaminants [Internet]. 1990;**7**(sup 1):S100-S102. Available from: http://www.tandfonline.com/doi/abs/10.1080/02652039009373857 [Cited: 10 May 2019]

[26] Casós K, Zaragozá MC, Zarkovic N, Zarkovic K, Andrisic L, Portero-Otín M, et al. Free Radical Research [Internet]. 2010;**44**(7):821-829. Available from: http://www.tandfonline.com/doi/full/10.3109/10715762.2010.485992 [Cited: 10 May 2019]

[27] Glemet HC, Ballantyne JS. Comparison of liver mitochondrial membranes from an agnathan (*Myxine glutinosa*), an elasmobranch (*Raja erinacea*) and a teleost fish (*Pleuronectes americanus*). Marine Biology [Internet]. 1996;**124**(4):509-518. Available from: http://link.springer.com/10.1007/BF00351032 [Cited: 10 May 2019]

[28] Gieseg S, Cuddihy S, Hill JV, Davison W. A comparison of plasma vitamin C and E levels in two Antarctic and two temperate water fish species. Comparative Biochemistry and Physiology - Part B [Internet]. 2000;**125**(3):371-378. Available from: https://www-sciencedirect-com.sire.ub.edu/science/article/pii/S0305049199001868 [Cited: 10 May 2019]

[29] Tappel ME, Chaudiere J, Tappel AL. Glutathione peroxidase activities of animal tissues. Comparative Biochemistry and Physiology - Part B [Internet]. 1982;**73**(4):945-949. Available from: http://www.ncbi.nlm.nih.gov/pubmed/7151426 [Cited: 10 May 2019]

[30] Mezes M. Investigations of vitamin E content and lipid peroxidation in blood and tissues of common carp (*Cyprinus carpio* L.). Aquacultura Hungarica. 1986;**5**:71-78

[31] Rudneva II. Blood antioxidant system of Black Sea elasmobranch and teleosts. Comparative Biochemistry and Physiology - Part C [Internet]. 1997;**118**(2):255-260. Available from: https://www.sciencedirect.com/science/article/pii/S0742841397001114 [Cited: 25 September 2018]

[32] Castro C, Peréz-Jiménez A, Coutinho F, Díaz-Rosales P, Serra CADR, Panserat S, et al. Dietary carbohydrate and lipid sources affect differently the oxidative status of European sea bass (*Dicentrarchus labrax*) juveniles. The British Journal of Nutrition. 2015;**114**(10):1584-1593

[33] Castro C, Peréz-Jiménez A, Coutinho F, Corraze G, Panserat S, Peres H, et al. Nutritional history does not modulate hepatic oxidative status of European sea bass (*Dicentrarchus labrax*) submitted to handling stress. Fish Physiology and Biochemistry. 2018;**44**(3):911-918

[34] Castro C, Diógenes AF, Coutinho F, Panserat S, Corraze G, Pérez-Jiménez A, et al. Liver and intestine oxidative status of gilthead sea bream fed vegetable oil and carbohydrate rich diets. Aquaculture. 2016;**464**:665-672

[35] Pérez-Jiménez A, Abellán E, Arizcun M, Cardenete G, Morales AE, Hidalgo MC. Dietary carbohydrates improve oxidative status of common dentex (*Dentex dentex*) juveniles, a carnivorous fish species. Comparative Biochemistry and Physiology – Part A. 2017;**203**:17-23. DOI: 10.1016/j.cbpa.2016.08.014

[36] Kokou F, Sarropoulou E, Cotou E, Rigos G, Henry M, Alexis M, et al. Effects of fish meal replacement by a soybean protein on growth, histology, selected immune and oxidative status markers of gilthead sea bream, *Sparus aurata*. Journal of the World Aquaculture Society. 2015;**46**(2):115-128

[37] Kokou F, Sarropoulou E, Cotou E, Kentouri M, Alexis M, Rigos G. Effects of graded dietary levels of soy protein concentrate supplemented with methionine and phosphate on the immune and antioxidant responses of gilthead sea bream (*Sparus aurata* L.). Fish and Shellfish Immunology [Internet]. 2017;**64**:111-121. DOI: 10.1016/j.fsi.2017.03.017

[38] Coutinho F, Castro C, Rufino-Palomares E, Ordonez-Grande B, Gallardo MA, Oliva-Teles A. Dietary glutamine supplementation effects on amino acid metabolism, intestinal nutrient absorption capacity and antioxidant response of gilthead sea bream (*Sparus aurata*) juveniles. Comp Biochem Physiol-Part A Mol Integr Physiol [Internet]. 2016;**191**:9-17. Available from: http://www.elsevier.com/inca/publications/store/5/2/5/4/6/4/%5Cnhttp://ovidsp.ovid.com/ovidweb.cgi?T=JS&PAGE=reference&D=emed18b&NEWS=N&AN=606263305

[39] Coutinho F, Castro C, Rufino-Palomares E, Ordóñez-Grande B, Gallardo MA, Kaushik S, et al. Dietary arginine surplus does not improve intestinal nutrient absorption capacity, amino acid metabolism and oxidative status of gilthead sea bream (*Sparus aurata*) juveniles. Aquaculture. 2016;**464**:480-488

[40] Feidantsis K, Kaitetzidou E, Mavrogiannis N, Michaelidis B, Kotzamanis Y, Antonopoulou E. Effect of taurine-enriched diets on the Hsp expression, MAPK activation and the antioxidant defence of the European sea bass (*Dicentrarchus labrax*). Aquaculture Nutrition. 2014;**20**(4):431-442

[41] Pérez-Jiménez A, Peres H, Rubio VC, Oliva-Teles A. The effect of hypoxia on intermediary metabolism and oxidative status in gilthead sea bream (*Sparus aurata*) fed on diets supplemented with methionine and white tea. Comparative Biochemistry and Physiology - Part C [Internet]. 2012;**155**(3):506-516. Available from: http://www.ncbi.nlm.nih.gov/pubmed/22227440

[42] Gisbert E, Skalli A, Campbell J, Solovyev MM, Rodríguez C, Dias J, et al. Spray-dried plasma promotes growth, modulates the activity of antioxidant defenses, and enhances the immune status of gilthead sea bream (*Sparus aurata*) fingerlings. Journal of Animal Science. 2015;**93**(1):278-286

[43] Esteban MA, Cordero H, Martínez-ToméM, Jiménez-Monreal AM, Bakhrouf A, Mahdhi A. Effect of dietary supplementation of probiotics and palm fruits extracts on the antioxidant enzyme gene expression in the mucosae of gilthead seabream (*Sparus aurata* L.). Fish & Shellfish Immunology. 2014;**39**(2):532-540

[44] Antonopoulou E, Kentepozidou E, Feidantsis K, Roufidou C, Despoti S, Chatzifotis S. Starvation and re-feeding affect Hsp expression, MAPK activation and antioxidant enzymes activity of European Sea Bass (*Dicentrarchus labrax*). Comparative Biochemistry and Physiology - Part A. 2013;**165**(1):79-88

[45] Sinha AK, AbdElgawad H, Zinta G, Dasan AF, Rasoloniriana R, Asard H, et al. Nutritional status as the key modulator of antioxidant responses induced by high environmental ammonia and salinity stress in European sea bass (*Dicentrarchus labrax*). PLoS One. 2015;**10**(8):1-29

[46] Lushchak VI, Bagnyukova TV. Effects of different environmental oxygen levels on free radical processes in fish. Comparative Biochemistry and Physiology - Part B. 2006;**144**(3):283-289

[47] Magnoni LJ, Martos-Sitcha JA, Queiroz A, Calduch-Giner JA, Gonçalves JFM, Rocha CMR, et al.

Dietary supplementation of heat-treated Gracilaria and Ulva seaweeds enhanced acute hypoxia tolerance in gilthead sea bream (*Sparus aurata*). Biology Open. 2017;**6**(6):897-908

[48] Feidantsis K, Pörtner H-O, Antonopoulou E, Michaelidis B. Synergistic effects of acute warming and low pH on cellular stress responses of the gilthead seabream *Sparus aurata*. Journal of Comparative Physiology B [Internet]. 2015;**185**(2):185-205. Available from: http://link.springer.com/10.1007/s00360-014-0875-3

[49] Elliot J. Tolerance and resistence to thermal stress in juvenile Atlantic salmon, *Salmo salar*. Freshwater Biology. 1991;**25**:61-70

[50] Ibarz A, Martín-Pérez M, Blasco J, BellidoD,deOliveiraE,Fernández-BorràsJ. Gilthead sea bream liver proteome altered at low temperatures by oxidative stress. Proteomics [Internet]. 2010;**10**(5):963-975. Available from: http://www.ncbi.nlm.nih.gov/pubmed/20131326 [Cited: 04 January 2018]

[51] Feidantsis K, Pörtner HO, Vlachonikola E, Antonopoulou E, Michaelidis B. Seasonal changes in metabolism and cellular stress phenomena in the Gilthead Sea Bream (*Sparus aurata*). Physiological and Biochemical Zoology [Internet]. 2018;**91**(3):878-895. Available from: http://www.journals.uchicago.edu/t-and-c [Cited: 13 July 2018]

[52] Vinagre C, Madeira D, Narciso L, Cabral HN, Diniz M. Effect of temperature on oxidative stress in fish: Lipid peroxidation and catalase activity in the muscle of juvenile seabass, *Dicentrarchus labrax*. Ecological Indicators. 2012;**23**:274-279

[53] Guerreiro I, Pérez-Jiménez A, Costas B, Oliva-Teles A. Effect of temperature and short chain fructooligosaccharides supplementation on the hepatic oxidative status and immune response of turbot (*Scophthalmus maximus*). Fish and Shellfish Immunology [Internet]. 2014;**40**(2):570-576. Available from: https://www-sciencedirect-com.sire.ub.edu/science/article/pii/S1050464814003027 [Cited: 10 May 2019]

[54] Birnie-Gauvin K, Costantini D, Cooke SJ, Willmore WG. A comparative and evolutionary approach to oxidative stress in fish: A review. Fish and Fisheries. 2017;**18**(5):928-942

[55] Stanca L, Petrache S, Serban A, Staicu A, Sima C, Munteanu M, et al. Interaction of silicon-based quantum dots with gibel carp liver: Oxidative and structural modifications. Nanoscale Research Letters [Internet]. 2013;**8**(1):254. Available from: http://nanoscalereslett.springeropen.com/articles/10.1186/1556-276X-8-254 [Cited: 04 January 2018]

[56] Hermenean A, Damache G, Albu P, Ardelean A, Ardelean G, Puiu Ardelean D, et al. Histopatological alterations and oxidative stress in liver and kidney of *Leuciscus cephalus* following exposure to heavy metals in the Tur River, North Western Romania. Ecotoxicology and Environmental Safety. 2015;**119**:198-205

[57] Esterbauer H, Cheeseman KH. Determination of aldehydic lipid peroxidation products: Malonaldehyde and 4-hydroxynonenal. Methods in Enzymology [Internet]. 1990;**186**:407-421. Available from: https://www.sciencedirect.com/science/article/pii/007668799086134H [Cited: 21 December 2018]

[58] Pryor WA, Porter NA. Suggested mechanisms for the production of 4-hydroxy-2-nonenal from the autoxidation of polyunsaturated fatty acids. Free Radical Biology & Medicine

[Internet]. 1990;**8**(6):541-543. Available from: https://www.sciencedirect.com/science/article/pii/089158499090153A [Cited: 21 December 2018]

[59] Chondrogianni N, Petropoulos I, Grimm S, Georgila K, Catalgol B, Friguet B, et al. Molecular aspects of medicine protein: Damage, repair and proteolysis. Molecular Aspects of Medicine [Internet]. 2014;**35**:1-71. DOI: 10.1016/j.mam.2012.09.001

[60] Carbonell T, Alva N, Sanchez-Nuño S, Dewey S, Gomes AV. Subnormothermic perfusion in the isolated rat liver preserves the antioxidant glutathione and enhances the function of the ubiquitin proteasome system. Oxidative Medicine and Cellular Longevity [Internet]. 2016;**2016**:1-12. Available from: https://www.hindawi.com/journals/omcl/2016/9324692/ [Cited: 25 September 2018]

[61] Kyprianou TD, Pörtner HO, Anestis A, Kostoglou B, Feidantsis K, Michaelidis B. Metabolic and molecular stress responses of gilthead seam bream *Sparus aurata* during exposure to low ambient temperature: An analysis of mechanisms underlying the winter syndrome. Journal of Comparative Physiology B: Biochemical, Systemic, and Environmental Physiology. 2010;**180**(7):1005-1018

[62] Feidantsis K, Antonopoulou E, Lazou A, Pörtner HO, Michaelidis B. Seasonal variations of cellular stress response of the gilthead sea bream (*Sparus aurata*). Journal of Comparative Physiology B: Biochemical, Systemic, and Environmental Physiology. 2013;**183**(5):625-639

[63] Hershko A, Ciechanover A. The ubiquitin system. Annual Review of Biochemistry [Internet]. 1998;**67**(1):425-479. Available from: http://www.annualreviews.org/doi/10.1146/annurev.biochem.67.1.425 [Cited: 25 September 2018]

[64] Jung T, Catalgol B, Grune T. The proteasomal system. Molecular Aspects of Medicine [Internet]. 2009;**30**(4):191-296. DOI: 10.1016/j.mam.2009.04.001

[65] Gogliettino M, Balestrieri M, Riccio A, Facchiano A, Fusco C, Palazzo VC, et al. Uncommon functional properties of the first piscine 26S proteasome from the Antarctic notothenioid *Trematomus bernacchii*. Bioscience Reports [Internet]. 2016;**36**(2):BSR20160022. Available from: http://www.ncbi.nlm.nih.gov/pubmed/26933238 [Cited: 25 September 2018]

[66] Kaushik S, Cuervo AM. Chaperone-mediated autophagy: A unique way to enter the lysosome world. Trends in Cell Biology [Internet]. 2012;**22**(8):407-417. Available from: https://www.sciencedirect.com/science/article/pii/S0962892412000931 [Cited: 25 September 2018]

[67] Salmerón C, Navarro I, Johnston IA, Gutiérrez J, Capilla E. Characterisation and expression analysis of cathepsins and ubiquitin-proteasome genes in gilthead sea bream (*Sparus aurata*) skeletal muscle. BMC Research Notes. 2015;**8**(1):1-15

[68] Weber D, Milkovic L, Bennett SJ, Griffiths HR, Zarkovic N, Grune T. Measurement of HNE-protein adducts in human plasma and serum by ELISA—Comparison of two primary antibodies. Redox Biology [Internet]. 2013;**1**(1):226-233. Available from: https://www.sciencedirect.com/science/article/pii/S2213231713000335 [Cited: 21 December 2018]